AQA

A-LEVEL

COURSEWORK WORKBOOK

Geography

Component 3: Geography fieldwork investigation (non-exam assessment)

David Holmes

HODDER
EDUCATION
LEARN MORE

My Coursework Planner

Section 1 Introduction

1.1 How to use this book

This book has been designed to help you to develop the skills necessary for success in your fieldwork investigation of the NEA – non-exam assessment for AQA A-level Geography. The book is divided into five main sections, each providing guidance on the different aspects of completing the investigation. Each section is made up of a series of topics. You will find guidance on what is required to achieve the top grades, followed by activities that will help to inform your thinking and practise some of the skills required for a quality geographical enquiry. Together these two strands of the book will take you through the different stages that are essential for success in your fieldwork investigation.

It is important to have a clear strategy for your coursework. This workbook will help you to achieve this in a planned and logical way. Use this book as the cornerstone of your research, **fieldwork** and writing – it is designed to be written in so don't be afraid to make use of all the space to refine your ideas and thinking. Feel free to scrawl all over it! It will also come in handy when reflecting on your progress, which forms the very last part of the enquiry process.

Section 1 introduces the enquiry process and the Pearson Edexcel mark scheme on which your investigation will be assessed.

Section 2 provides important guidance on how to focus and shape an idea into a tighter theme that can be successfully investigated. Section 3 is made up of a number of activities that will help you understand the significance of geographical sources, including advice on using published literature as a source of ideas for your own design and methodologies.

Sections 4 and 5 focus on the mechanics of fieldwork design and delivery, as well as writing up your findings. Section 5 also has extracts for you to review and evaluate as well as ideas for your own investigation. Work through all of these as they have been designed for you to practise and develop different skills required for the latter stages of enquiry.

At the back of this book you will find several logs to record your progress. While the 'My progress' boxes throughout this workbook give you a chance to apply what you have learnt from an activity to your own investigation and to record some of your thinking, the logs at the back will serve as a working document that you should be adding to from day one.

There is also a glossary of key terms and technical vocabulary. These are indicated in the text in green bold.

1.2 The enquiry process and key questions answered

What is enquiry?

Enquiry can be thought of as a series of small stages or steps in an overall process to try to find an answer to a question or hypothesis that has been set. A clear understanding of this sequence is important since it forms the basis and structure of the AQA fieldwork investigation mark scheme (see pages 7–11). Geographical enquiry is important since it:

- creates curiosity and allows you to challenge concepts, models, theories and beliefs

- encourages an evidence-based approach to using research and other source materials
- allows a way of thinking that is often more reflective and that can help with understanding the complexity of 'messy', real-world geography.

Stages in a geographical enquiry process are shown in Table 1. The AQA investigation pathway closely matches this, but uses slightly different terminology.

Table 1

	Stage	Description
1	Purpose, identification of a suitable question/aim/hypothesis and developing a focus	• Identify appropriate field research questions/aims/hypotheses, based on knowledge and understanding of relevant aspects of physical and/or human geography. • Research the relevant literature sources linked to possible fieldwork opportunities presented by the environment, considering their practicality and relationship to compulsory and optional content. • Understand the nature of the current literature research relevant to the focus. This should be clearly and appropriately referenced within the written report.
2	Designing the fieldwork methodologies, research and selection of appropriate equipment	• Think about how to observe and record geographical ideas in the field, and to design appropriate data-collection strategies, taking account of sampling and the frequency and timing of observation. • Select practical field methodologies (primary) appropriate to the investigation (may include a combination of qualitative and quantitative techniques).
3	Data collation and presentation	• Use appropriate diagrams, graphs and maps, and, if relevant, geospatial technologies, to select and present relevant aspects of the investigation outcomes.
4	Analysis, interpretation and explanation of results and information	• Use appropriate techniques for analysing field data and research information. • Provide a coherent analysis of fieldwork findings and results linked to a specific geographical focus.
5	Conclusions and critical reflection on methods and results	• Use knowledge and understanding to question and interpret meaning from the investigation (theory, concepts, comparisons), through the significance of conclusions. • Demonstrate the ability to interrogate and critically examine field data (including any measurement errors) in order to comment on its accuracy and/or the extent to which it is both representative and reliable.
6	Recognising the wider geographical context	• Explain how the results relate to the wider geographical context and use the experience to extend geographical understanding. • Show an understanding of the ethical dimensions of field research.

Aspects of enquiry

Motivations for a geographical enquiry can have many components. You should give some thought to all of these ideas when at the initial planning stages of the enquiry process (see Figure 1).

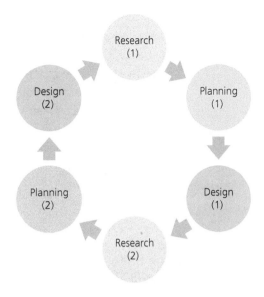

Figure 2 The reflective nature of research.

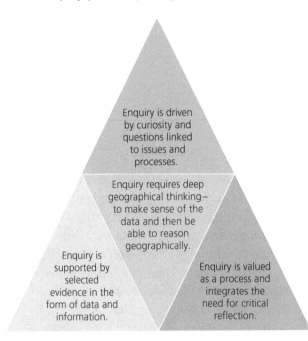

Figure 1 The components of enquiry.

Enquiry should be reflective

The stages in an enquiry should be seen as a reflective and iterative process. Sometimes there is a need to revisit and rethink aspects of what you have done. This may perhaps drive the rest of the enquiry in a slightly different direction from originally intended (see Figure 2).

What does the fieldwork investigation assess?

The fieldwork investigation is worth 20 per cent of your A-level grade. It is really a test of your ability to undertake an enquiry on your own. It focuses on the collection and interpretation of different types of data, therefore you will need to analyse and evaluate in a geographical manner.

What can the fieldwork investigation be about?

The coursework can be based on either physical or human geography, or a combination of both. Whatever you choose, there should be a clear link to the specification and it must be geographical (rather than say historical) in its focus, see Table 2.

Table 2 Linking the Investigation to the specification.

3.1.3.1 Coasts as natural systems

Systems in physical geography: systems concepts and their application to the development of coastal landscapes – inputs, outputs, energy, stores/components, flows/transfers, positive/negative feedback, dynamic equilibrium. The concepts of landform and landscape and how related landforms combine to form characteristic landscapes.

3.1.3.2 Systems and processes

Sources of energy in coastal environments: winds, waves (constructive and destructive), currents and tides. Low energy and high energy coasts.

Sediment sources, cells and budges.

Geomorphological processes: weathering, mass movement, erosion, transportation and deposition.

Distinctively coastal processes: marine: erosion – hydraulic action, wave quarrying, corrosion/abrasion, cavitation, solution, attrition; transportation: traction, suspension (longshore/littoral drift) and deposition; sub-aerial weathering, mass movement and runoff.

You could look at sediment size and distribution on a beach that would link to 'transport' in 3.1.3.2 for instance.

How much help can I get?

This is an independent study, so you cannot get specific help from your teacher on the writing of your essay. Your teacher is allowed to discuss with you the focus for your enquiry, but cannot give you a title or direct feedback on your proposals. After that you are pretty much on your own. The way that the coursework will be marked is explained on pages 7–11; however, the main thing to remember is that the *writing-up must be your own work*.

How long is the fieldwork investigation?

Your investigation should be 3000–4000 words in length. If you write less than 3000 words you would be unable to deal with the topic in sufficient depth. There is no direct penalty for going over-length, but it may mean that you produce a less concise piece of work, which ultimately could be detrimental to your grade.

What are the essential components of the fieldwork investigation?

- It must be based on a question or issue defined and developed by you individually to address an aim, question and/or hypothesis relating to any of the compulsory or optional content in the course.
- The work needs to incorporate field data and/ or evidence from field investigations, collected individually or in groups
- It must be based around your own research, including primary data and, if relevant, secondary data (sourced by you).
- You are responsible for the context, analysis and summary of findings from data and evidence.

Can I have the same title as someone else?

AQA's guidance says that titles can be the same or similar as long as the student has developed the title independently. Depending on how your fieldwork is organised you may find that there is some overlap between your approach and that of other students. Try to get individuality through your approach, such as different maps, equipment, methodology, design, survey areas, and so on.

How much data needs to be collected?

The amount of data to be collected is very much dependent on the focus of your investigation.

Depending on what you have chosen you will need to balance primary, secondary, qualitative and quantitative data, and recognise what value these bring to an investigation. In choosing your data collection techniques at the planning stages of your investigation, you should be able to explain why you have chosen them, as well as how this will support the later analysis.

You may be able to collect your data in one day, although think about whether it might be a good idea to repeat the survey at different times, or separate the data collection times so that a temporal pattern of change can be investigated.

How do I find out how to present and analyse my data?

You will probably have some support from your teacher during your geography course before starting your independent investigation. They will show you general approaches to data presentation and analysis, although these won't be specific to your particular focus. There are lots of other written and online sources. YouTube Kahn Academy www.youtube.com/khanacademy, for example, can be very helpful for showing statistical analysis and procedures.

What should be included when submitting the fieldwork investigation?

- The fieldwork investigation itself, which must include a bibliography (normally ordered alphabetically by author's surname) of the works referred to in the coursework.
- The fieldwork investigation proposal form.
- An investigation report authentication sheet, which must be signed by you as well as your teacher.
- You are also allowed to submit an appendix with any extended extracts, for example copies of interviews, transcripts or large sections of secondary information, which you have referred to in the main part of the coursework.

How is the fieldwork investigation marked?

The enquiry is initially marked by your teacher and, if your school or college has a large number of students studying Geography, it may need to be moderated internally by another teacher. All schools and colleges then have to submit a sample to the awarding body to be moderated by an external moderator a month or so before your final exams.

1.3 Demystifying the AQA mark scheme

The mark scheme is a set of descriptive criteria upon which the fieldwork investigation is marked by a teacher. In order to do well you need to understand the specific terminology and phrases in the mark scheme.

There are different marks allocated for different Assessment Objectives (AOs). The breakdown for the AQA NEA is shown in Table 3. Table 4 shows how many words might be written for each section, assuming a 3000–4000 words piece of coursework.

In order to understand the mark scheme, on the next few pages we will look at the top bands in each section. The important words or phrases are highlighted and annotated to help you understand what is required to achieve the higher marks.

Table 3 The Assessment Objectives.

	Description	% for NEA
AO1	Demonstrate knowledge and understanding of places, environments, concepts, processes, interactions and change, at a variety of scales.	0%
AO2	Apply knowledge and understanding in different contexts to interpret, analyse and evaluate geographical information and issues.	10%
AO3	Use a variety of relevant quantitative, qualitative and fieldwork skills to: <ul><li>investigate geographical questions and issues</li><li>interpret, analyse and evaluate data and evidence</li><li>construct arguments and draw conclusions.</li></ul>	90%

Table 4 Areas and key ideas for the AQA fieldwork investigation (NEA) and possible number of words in each component.

	Area	Key extracts from the AQA generic mark scheme
1	**Introduction and preliminary research** (10 marks) 400–600 words	<ul><li>Accurate and relevant geographical knowledge and understanding of location, geographical theory and comparative context.</li><li>Uses a range of relevant literature geographical sources in order to identify/obtain accurate geographical information and data that support the investigation.</li><li>Defines a research question that provides an appropriate framework for investigation at a manageable scale.</li></ul>
2	**Methods of field investigation** (15 marks) 300–500 words	<ul><li>Chooses appropriate methods to collect a record of phenomena in the field.</li><li>Designs a valid sampling framework explicitly linked and appropriate to the geographical focus.</li><li>Considers timing and frequency of observations.</li><li>Uses methods with high levels of accuracy/precision to obtain reliable data.</li></ul>
3	**Methods of critical analysis** (20 marks) 1300–1700 words	<ul><li>Uses appropriate diagrams, graphs and maps, using technologies to select and present relevant aspects of the investigation outcomes.</li><li>Uses geographical skills to analyse data in order to show evidenced connections and accurate statistical/geographical significance of data.</li><li>Critically examines field observation data (including any measurement errors) in order to comment on its accuracy and/or the extent to which it is representative.</li><li>Relates concepts and theory to own field observations.</li></ul>
4	**Conclusions, evaluation and presentation** (15 marks) 1000–1300 words	<ul><li>Shows relevant links between the investigation's conclusions and a broader geographical context.</li><li>Shows an understanding of the ethical dimension of field research.</li><li>Synthesises research findings coherently.</li><li>Considers the reliability of evidence and validity of conclusions.</li><li>Writes up results clearly and logically and uses appropriate geographical terminology.</li><li>Conclusions use evidence and link to concepts/theory and make a well-argued case.</li></ul>

Level 4: 9–10 marks

A research question(s) is effectively identified and is completely referenced to the specification. (AO3)

Well-supported by thorough use of relevant literature sources. (AO3)

Theoretical and comparative contexts are well-understood and well-stated. (AO3)

This is asking you to link your idea or topic into an area of the AQA specification. The link should be explicit and obvious.

Geographical theory is likely something you have found in your literature search and may have come from a magazine, book, journal or other technical document. A theory is really a testable assumption which you can use to match with your own fieldwork data. You may find the two disagree.

A research question is the focus for what you are studying. It's an overarching idea that it may be broken down into sub-questions, aims or hypotheses but all of these must directly link to the bigger research question.

Literature sources are from reading and research that you have done. They might help you understand more about a place and geographical processes. They might also help you make connections and wider links outside of the area you are studying. For example, you might be studying the identity of a local place because of change in different town centres that you have read about.

Your sources should allow you to link to other areas or locations where similar studies or pieces of research have been undertaken.

You need to show that you can join together concepts that underpin the geography of the research question. For example, if you are measuring a beach gradient, what are the models and theories that help us to understand the processes at work, in different places and at different times? How do those processes affect and impact on the beach gradient?

My progress

Having reviewed Area 1 of the mark scheme, complete the checklist to show your understanding of this area and its terminology. Use a tick or a question mark, filling in any gaps in understanding as necessary.

Introduction and preliminary research – Checklist	✔
Manageable scale so that the research question can be answered	
Research information is used to construct a research question	
Uses different literature sources	
Appropriate framework linked to the specification	
Shows a relevant comparative context	
Demonstrates knowledge of both location and relevant theory and concepts	

Interpreting Area 2: Methods of field investigation (15 marks)

AO3: 15 marks

Level 4: 12–15 marks

Detailed use of a range of appropriate observational, recording and other data collection approaches including sampling. (AO3)

Thorough and well-reasoned justification of data collection approaches. (AO3)

This might relate to how many times (possibly repeats) you carry out your observations. Generally, more repeats mean greater reliability. Ethical dimensions (see page 63) will also need consideration for both working in communities and in natural landscapes.

Detailed demonstration of practical knowledge and understanding of field methodologies appropriate to the investigation of human and physical processes. (AO3)

This is about making sure that your data collection is well matched to your research focus. Any techniques that are used should link precisely to your title and should not be unnecessary fillers. Once again, there is the need to think about timing and frequency of observations as well as the risk assessment.

Detailed implementation of chosen methodologies to collect data/information of good quality and relevant to the topic under investigation. (AO3)

This will vary by project, but it's important to include both primary and secondary information so that it gives scope for discussion and analysis. Too much of one type of data on its own may create problems.

A pilot survey, as detailed on page 34, is like a fieldwork 'practice run'. So, techniques can be tried out to ensure that they produce good quality data and information which is relevant to your research focus.

This is a consideration of which sampling design (size, spacing) is the most appropriate for the data being collected. It really links with the next point about timing and frequency.

Justification might include considerations of timing, access, safety and secondary data availability. Timing is when you collect data or make observations which may need justification. This will have a big impact on your findings since most places change from minute to minute and day to day. For some types of projects this will be less important; for most however, it will be a significant consideration (and is often overlooked).

My progress

Having reviewed Area 2 of the mark scheme, complete the checklist to show your understanding of this area and its terminology. As before, use a tick or a question mark, filling in any gaps in understanding as necessary.

Methods of field investigation – Checklist	✔
Design of a valid sampling framework	
Frequency and timing of observations	
Consideration of the ethical dimensions	
Range of data (that is linked to the research focus)	
Risk assessment	
Data availability and manageability	
Reliability	

Interpreting Area 3: Methods of critical analysis (20 marks)

AO2: 6 marks

AO3: 14 marks

This is about making sense of what you have found. Remember that sample size and location will influence this and your choice of statistical tools for instance, to help see what the geography is showing.

Level 3: 15–20 marks

Effective demonstration of knowledge and understanding of the techniques appropriate for analysing field data and information and for representing results.

Careful selection of data presentation techniques is needed to show patterns, trends and anomalies. This may vary by location so data representation at a spatial scale is important, i.e. allowed the reader to compare between locations or sites. Don't just think there will be reward for complex or 'fancy' graphical procedures.

Here you need to try to show links between different aspects of the data. Use of a spider diagram, for instance, might help.

Different types of data and information will require different tools.

Thorough ability to select suitable quantitative or qualitative approaches and to apply them.

This is evaluation of techniques (not just limitations). Did the methods produce accurate and/ or reliable results? Did your design and methods produce data that might have been anomalous or introduce bias?

Accuracy is covered later in this workbook, but you should think about it as being how much your data allows you to make judgements around whether you have really found out the answer that you set out to find out about.

Thorough ability to interrogate and critically examine field data in order to comment on its accuracy and/or the extent to which it is representative.

Representative means how closely the characteristics of the sample match the characteristics of the population.

Complete use of the experience to extend geographical understanding.

Use evidence (data/ facts) obtained from your research and presentation and integrate them into your conclusions.

At this stage you might want to make reference to specialised concept links, for example risk or resilience, or perhaps appreciating that your study forms part of a wider and more complex geographical system. It's really trying to link into why your study might be of interest to perhaps a local resident or other interested person or group.

Effective application of existing knowledge, theory and concepts to order and understand field observations.

You will need to link together the discoveries from your own primary data with published information that forms part of your literature review. This could include other secondary data. There may be differences and similarities that can be commented upon and written about. Here you are trying to show an appreciation of geographical connections and processes.

My progress

Having reviewed Area 3 of the mark scheme, complete the checklist to show your understanding of this area and its terminology. As before, use a tick or a question mark, filling in any gaps in understanding as necessary.

Methods of critical analysis – Checklist	✔
Statistical skills (if appropriate to the data collected)	
Making geographical connections using evidence	
Clear and technically accurate presentation	
Accuracy and representative	
Makes sense of findings	
Evidenced based links between your data and research concepts	

This whole point is linked to the quality of written communication and clarity of geographical expression. A danger is that many students are not concise and simply write too much and repeat themselves.

This is not only evaluating the reliability of the evidence but also the accuracy and representativeness of the data or evidence.

These are high-level synthesis skills – bringing together what you have found and what you know into a clear summary. You will also be expected to then link these ideas to your theory/concepts and the wider geographical understanding. Don't be worried if your conclusion doesn't turn out as expected.

Interpreting Area 4: Conclusions, evaluation and presentation (15 marks)

AO3: 15 marks

Level 3: 12–15 marks

Thorough ability to write up field results clearly and logically, using a range of presentation methods.

Effective evaluation and reflection on the fieldwork investigation.

This is whether your quantitative or qualitative analysis can be trusted. You need to refer back to sample size, frequency, timing and so on, as well as particular methods.

Complete explanation of how the results relate to the wider context(s).

These are considered in much more detail in Section 5. 'Ethical dimensions' are essentially about respecting other people or environments and safeguarding data and information that might be obtained.

Thorough understanding of the ethical dimensions of field research.

Thorough ability to write a coherent analysis of fieldwork findings in order to answer a specific geographical question.

This whole point is linked to quality of written communication and clarity of geographical expression. Again, correct use of geographical terminology and integration with ideas from the literature research adds refinement.

Draws effectively on evidence and theory to make a well-argued case.

Here you are building an argument – a series of evidenced conclusions leading to an overall judgement (argument) that returns to the original hypothesis or question and indicates the degree to which the evidence collected provides answers.

My progress

Having reviewed the final area of the mark scheme, complete the checklist to show your understanding of this area and its terminology. As before, use a tick or a question mark, filling in any gaps in understanding as necessary.

Conclusions, evaluation and presentation – Checklist	✔
Accurate and relevant geographical knowledge of both location and geographical theory including comparative context	
Work is well ordered and grammar is sound	
Relevant links between the conclusions and a wider geographical context	
Evaluation of the reliability of the evidence and the validity of the conclusions	
Synthesises research findings to produce convincing conclusions which are fully supported by relevant evidence and concepts	

Section 2 Making a start

2.1 Choosing a topic and focus

The topic for your NEA must be your own choice. You must come up with the idea and focus independently. It's an important part of your A-level Geography, remember that it's nearly worth 20 per cent, so the topic does need some thought.

Ideally, and in no particular order:

It should be interesting

As it's worth 60 marks overall, you really need to be thinking of spending perhaps 30–40 hours writing up after you have completed your fieldwork. That will of course be something that your teacher or tutor will advise you on. Given the amount of time you will need to spend on it, it's important to find a topic that interests you:

- It might be something or somewhere you have studied before, but it is likely you will need to go into greater depth or complexity or approach it from a different perspective.
- It might be something you have seen or read about and want to find out more about.
- It might be something that affects your family or the local community and is causing controversy.
- It might touch on other areas of your study or hobbies that you enjoy such as art, music, photography, business, sport and economics.
- It might be a chance to do something contemporary, for example using social media or blogging to approach a socially-derived idea.
- It might be something to show university admissions tutors that you are interested in different aspects of the subject and are not afraid of tackling new topics.
- It might be that you like a familiar topic or environment, but are prepared to look at it with a fresh pair of eyes or using a different methodology, for example.

It should be manageable and deliver the requirements of the mark scheme

- It shouldn't be such a big topic that you can't really control either the research or fieldwork.

- It should be based at a spatial scale that is both manageable (usually small-scale) and appropriate to the topic.
- It should be a topic that allows you to find a range of readily accessible primary and secondary data/information which allows you to reach a substantiated conclusion.
- It should be planned so that the area being investigated is both accessible and safe to work in.

It should be geographical

- The topic should allow you to challenge a concept, model, theory, idea or even an assumption or belief.
- It should be on a topic which can be identified within the specification (you will have to show this on the fieldwork investigation proposal form – see page 18).
- It can be based on a topic which spans several areas of the specification.
- It should allow you to link to a 'bigger picture' idea or wider physical or human system.

It should be answerable

- It should be a topic or idea that is likely to be answerable within the recommended 3000–4000 words.
- It might be that you don't know the answer before you start, or that the answer at the end is not the one that is expected.
- It should be a topic that allows you to reach conclusions that might be partial, tentative or even incomplete.
- However, it should not be answerable with a simple 'yes' or 'no'.
- It should also avoid a truism, for example 'Does the pedestrian flow increase towards the centre of the town?'

Mind map

Here is an example of a mind map created by a student who is considering the focus of their NEA. They have thought about their own skills and talents as well as priorities when considering a possible idea and topic for their investigation. They have begun by writing them down in a mind map, and have gone on to number them to create a rank order of importance.

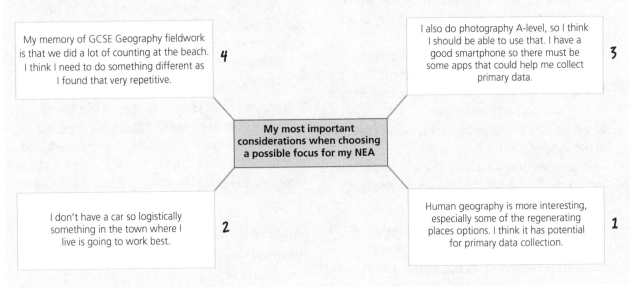

My memory of GCSE Geography fieldwork is that we did a lot of counting at the beach. I think I need to do something different as I found that very repetitive. **4**

I also do photography A-level, so I think I should be able to use that. I have a good smartphone so there must be some apps that could help me collect primary data. **3**

My most important considerations when choosing a possible focus for my NEA

I don't have a car so logistically something in the town where I live is going to work best. **2**

Human geography is more interesting, especially some of the regenerating places options. I think it has potential for primary data collection. **1**

My progress

Use the model above to set out your own priorities. Start by thinking of factors that are important for you.

Some important considerations for me are:

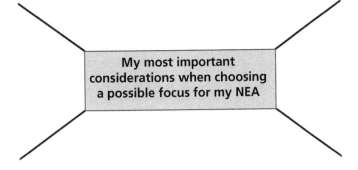

My most important considerations when choosing a possible focus for my NEA

Now number them in order of importance to you.

2.2 Initial background research

By this point you may have a very broad and non-specific idea of an area of geography that you might want to begin to investigate. You are unlikely to know at this point, however, where to, or even how to, approach the investigation. So, the initial background research you need to do at this point may serve several overlapping objectives.

- To read-up about sources of ideas that can be developed into a more appropriate focus.
- To establish what research and information might be available on particular topics.
- To get an idea around a topic or concept that links to the specification.
- To find out more about a local area and topical issues.
- To find out how a possible idea might link to a wider topic or system.

It is important to make notes or take photographs of sources used at this point so that you can begin to compile a list of references. At this stage it doesn't need to be formalised; instead just keep a record that you can use later to complete a more finalised bibliography.

Where to look

We will cover more about the literature research and review as well as sources of information in Section 3, page 20. However, at this point in the enquiry process there are likely to be a few avenues for the initial research that you might explore.

Local newspapers and forums

These are good starting points especially for thinking about local issues (or conflicting views) in an area.

Technical reports

These are available on the internet often as pdf files, but can be very useful if they match your topic area. Good for concepts and ideas.

Magazines

Publications such as *Geography Review* and *GeoFactsheet* have searchable archives. These may provide ideas for manageable questions.

Importantly, such sources may give you initial ideas around a concept or question, as well as a starting point with regard to data collection approaches. Your textbook and school notes will also provide some information, but be aware that these will be generalised and may not help you with creating a specific focus at this stage. They will also lack any local dimension.

Different reading methods

It is often said that you should begin your research by reading, and this is certainly good advice. However, at this stage in the enquiry process you are likely to need to be either skimming or scanning to cover a quite a large amount of material:

- Skimming – basic quick reading to determine the quality of the information.
- Scanning – quick reading to locate key words or phrases.

This may be from sources that you have found online, or printed copies of information that are available in a library. If possible, use key words and tags to locate the information quickly. Again, you are reminded to keep a record of any sources.

Keeping up to date – local sources of information

Wherever you intend to undertake your fieldwork investigation it is important to 'connect' the geography at a local scale. In other words, look for a local angle (perhaps a local issue related to planning, pollution, traffic, the high street, coastal erosion, and so on). It is often called a 'local media scan'. Local issues can of course be both physical and/or human topics.

Figure 3 includes some examples of local online sources that might help to create an idea around affordable housing, changing communities and gentrification.

Exciting times ahead for Lewisham with proposed tube line extension?

Plans to extend the Bakerloo line could regenerate Lewisham as the extended East London line did Hoxton and Dalston. As Lewisham currently relies on buses and over ground trains, extending the tube could vastly improve local transport. The news has already triggered large investment in the area by housebuilders. Transport for London says the project, supported by Mayor Sadiq Khan, will cost £3.6 billion. If approved, it could start in 2023 and services should start running by 2028.

'Rapid Gentrification' cause of Lewisham residents' disposable incomes rising fastest in UK

Adding Lewisham to the tube map doesn't guarantee regeneration

Housing developers are rushing in on the back of £3.6bn project to make Lewisham part of the Bakerloo line. Will this regenerate the area or is a more holistic approach needed for a lasting impact?

Figure 3

My progress

Choose any three of the ideas you have identified as part of your local media scan.

The three most important themes or local issues for me are:

1 _____

2 _____

3 _____

Use this box to keep a written record of your sources of information.

Try to evaluate each of your broad themes by thinking about an advantage and disadvantage for each.

Theme 1:

Theme 2:

Theme 3:

You can also refer to the research logs at the end of this workbook (pages 73–77).

2.3 Generating a title and developing aims, questions or hypotheses

So, at this point you have got a fairly good idea of what you might want to do. You know broadly which area of geography you are focused on, and where you might undertake the fieldwork. You have also done some initial research and fact-finding to check that your idea is feasible. The next job is to refine your initial research into an appropriate title. This can be modified and re-worded at a later date, but for now it helps to have a clear focus, a 'working title', at least.

Writing your title is very important as the NEA will, in part, be marked according to how well you have answered your own question or addressed the focus in the title. This is very advantageous in one way as you decide the title (like writing your own exam question!) but it does mean you need to carefully word the question so that you can focus on answering it.

The title: a question or a statement?

Titles generally take the form of either:

1 an issue

2 an aim

3 a testable statement – hypothesis

4 a research question.

The choice is obviously yours, but there are either:

- those that focus on *spatial/areal* or *temporal* differences, or
- those that focus on relationships between *variables*.

Table 5 shows how the title on the same topic can vary depending on whether it is an aim, a hypothesis or a question.

Table 5

Issue (statement)	An analysis of beach use and conflicts between different visitor 'segments' in the coastal resort of Sitges, Barcelona.
Aim	To study the changes in quality of life between two different coastal suburbs in the area of Sitges, Barcelona.
Hypothesis (testable statement)	The shingle beaches at Sitges (Barcelona) demonstrate a positive relationship between gradient and pebble size.
(Research) Question	To what extent does land use of the coastal zone impact on the choice of management strategies used in Sitges, Barcelona?

Is there geography and a location?

Geographical context – the title you choose must have a clear identifiable link to the specification (you will need to include this in the proposal form). Often the 'geography' part comes from a key word or phrase, linked to a theory or concept. It could also be linked to the specialised concepts.

AQA specialised (synoptic) concepts: causality, system, feedback, inequality, identity, globalisation, interdependence, mitigation and adaptation, sustainability, risk, resilience and thresholds.

Location context – the title must have a clear and precise spatial dimension to it. Rather than just giving a city, town or broad area, try to narrow down the area to a particular place or even a census output area or ward.

An example of this would be: How does gender influence perception of place in Bristol? This can be better localised to:

How does gender influence perception of place in Stokes Croft and St Paul's, Bristol?

Developing sub-questions?

Sub-questions, or key questions (as they are sometimes known) are often used in fieldwork investigations. They often provide a framework and help you break down a larger title or focus into more manageable parts. Each sub-question must be answerable and directly linked to the title.

There is a danger that sub-questions can cause the investigation to drift away from its original focus. Table 6 shows how sub-questions can easily drift and make the project much bigger than originally intended. In this instance, 'gentrification' needs to be narrowed to something workable, manageable and answerable.

Table 6

Original title	Sub-questions
To what extent has gentrification in Shrewsbury impacted on the local community?	1. Has gentrification increased local employment? 2. Has gentrification reduced deprivation? 3. Has gentrification increased transport congestion? 4. Has gentrification improved the quality of the environment? 5. Has gentrification caused a decrease in local crime rates?

1 Look at these titles and complete the table by deciding whether or not you think they are suitable for a fieldwork investigation.

Title	Suitable	Not suitable
What are the economic benefits of the Emirates stadium for local businesses?		
Why are the groynes at Porlock Bay not maintained?		
To investigate the processes responsible for a changing coastline along a section of the Llyn Peninsular.		
To what extent has the regeneration of Telford's Southwater region been successful?		
To what extent has regeneration changed people's quality of life in Wembley Central?		
How does beach profile and sediment size vary according to implemented management strategies at Newhaven?		
What is the evidence for glaciation isolated from the Welsh ice sheet in parts of Snowdonia, North Wales?		
The relationship between increased urbanisation, changing climatic factors and flood risk in two contrasting areas within the Lake District.		
How does social media in El Raval affect sense of place for tourists?		

2 What criteria are you using to judge the suitability of your own title or focus?

My progress

Has doing this exercise affected the way that you think about your own title? Help shape your thinking by filling out the table below.

Topic and focus	Title
What I **first** thought about my idea:	What I **first** thought was a good title:
What I **now** think about my idea:	What I **now** think would be a good title:

You can also refer to the research logs at the end of this workbook (pages 73–77).

2.4 Fieldwork investigation proposal form: initial draft

NEA proposal
To be completed by the candidate

Investigation title: *This is a work in progress and can easily be adapted at a later date if necessary.*

How the title links to the specification content

Look for a relevant specification match, e.g. 3.1.3.1 Coasts as natural systems. Use some of the words from the specification to show the link explicitly, e.g. inputs, dynamic equilibrium.

Planned investigation hypothesis or question/sub-questions

Note – planned investigation. Written at the time of the development of a title and research focus. Be careful with too many sub-questions 'diluting' and extending the original focus.

Investigation focus – indication of how the enquiry will enable the candidate to address the investigation title and explore the theme in relation to the chosen geographical area

This is the broad context and outline of the enquiry. Your teacher can give you general advice and feedback as part of a planning conversation.

Planned methodology – indication of qualitative and/or quantitative techniques including primary and, if relevant, secondary data collection techniques. Indication of the planned sampling strategy or strategies

To start with you may only have a few ideas that will be refined as you get more into the research and developing knowledge around the enquiry. Again, this section can be modified retrospectively, e.g. after a pilot survey.

Data collection: ☐ Individual ☐ Group

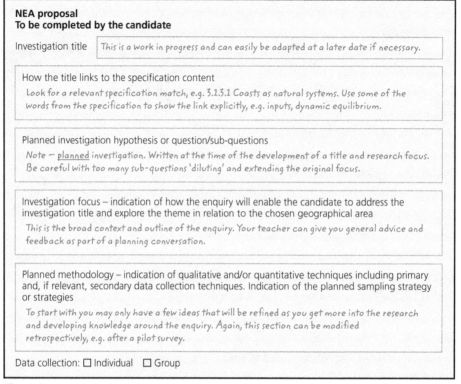

Figure 4

The proposal form is a planning document. It allows you to set out your ideas in a structured way so that it can be checked and approved by your teacher(s) or supervisor.

Figure 4 is an example of an extract from a blank AQA candidate record and proposal form. The blue font offers some simple guidance on how to complete the form. If you are unsure, ask your teacher to show you examples of completed forms and commentaries, from AQA.

The completed form is submitted to AQA as part of the moderation process.

Teachers are not allowed to give you specific guidance on your project, but they are permitted to give support on the enquiry process in the early stages. Going to your teacher with a well thought out proposal form will mean that there is a chance of appropriate feedback.

Figure 5 show examples of teacher feedback as part of the planning phase. Your teacher is able to support you with general guidance and feedback, but cannot offer specific information.

Note that your teacher can provide information about **risk assessments** and ethical considerations. This is to make sure that you conduct your investigation in a safe and appropriate manner.

You can revisit, rework and adapt the proposal form at any point. This might be especially relevant after completing a test or **pilot survey** which may lead you to re-evaluate your focus or modify your fieldwork design. A good suggestion is to use a different colour for any modifications and to initial and date the changes that you make.

Context 1: A study of glacial landscape features and morphology

Teacher guidance and feedback:

I think you need to reconsider your proposed data analysis. Importantly you need to consider both the sample size and methodology (for example, equipment) that we have covered in class at the beginning of this term. I'm not convinced that what you have suggested on the form will actually answer the question you have set out.

Context 2: Does town XX have a successful town centre?

Teacher guidance and feedback:

It's going to be critical to establish a clear definition and meaning of 'successful'. You will probably need to do more research as to how other people might both define and measure this kind of criteria. If you find something workable then adapt the approach for your own use.

Figure 5

Reviewing a proposal form extract

Review the extract from a student's planning sheet, before they completed their proposal form, shown below. What advice could you offer to them on headings 1–4 in the proposal form? Work with another student to share your thoughts.

---

(1) Proposed title

> How does the environmental quality vary within xxx (a medium-sized town)?

(2) Key questions/hypotheses

> 1. How does the environment vary at different locations within xxx?
> 2. How does the use of land (buildings, building type) affect the environment within xxx?
> 3. To what extent do pollution levels within xxx have a negative effect on the environment?
> 4. What strategies are there to make xxx more sustainable?

(3) Investigation focus and context

> xxx is a constantly changing rural area, meaning that the pollution levels, congestion levels and more are also changing. This then results in the environment of the area never being constant due to many different players and factors. By looking at a range of different ideas, the different factors affecting the environment, and whether they are having a positive or negative impact, can be determined.

(4) Proposed methods of data collection

Primary data:
- Land use survey, photographs, questionnaires and interviews, noise pollution survey, environmental quality survey

Secondary data:
- GIS comparison of historical maps and satellite imagery, council data, air pollution online collection

---

My progress

Here is a chance to review your ideas so far in sections 2.1–2.3. Complete the checklist below, correcting any 'No' or 'Not sure' responses, before completing your own proposal form. There is also space for you to add your own items to the checklist.

	Yes	Not sure	No
My title can be **realistically answered** and is **manageable**.			
The focus is not too large. It is at an **appropriate scale**.			
I am focusing my NEA around a **concept** or **geographical idea** linked to an aspect of the **specification**.			
I can **show individuality** even if I'm collecting data in an area with other students.			
I am going to work in a place that is both **safe** and **accessible**.			
There is data which are known to be available **before** embarking on the study.			
I understand the potential for **design** and **sampling procedures** as part of the planned methodology (including adapting pre-existing studies).			
I can modify or **create my own recording sheets** matched to the focus.			
There is an appropriate source of **accessible secondary data**.			
It is likely that conclusions can be **drawn** and will be **focused**.			
The focus will allow discussion about **reliability** and **accuracy**.			
The methods and techniques are suitable for the investigation focus.			

Section 3 Research and literature sources

3.1 Understanding the different types of research and literature sources

The literature sources are available to help you to develop a good understanding of, and insight into, previous research. Put simply, a literature search or review involves finding the current information on a particular topic. You have already been introduced to the idea of background research on page 14.

Research to build the conceptual framework

One of the reasons to start the literature search and review is to help you to develop a conceptual framework which helps you to understand your chosen topic more clearly. One way this might be done is by using a concept map to show ideas (or concepts) and the relationship between ideas. By reading the research you can begin to develop your own initial concept or mind map. This could, of course, form part of your introduction. Your research helps you to achieve this.

Table 7

Identify concepts	These are the building blocks of your investigation, i.e. the variables being investigated.
Define concepts	Concepts must be clearly defined for the research. Dictionaries may be helpful, but sometimes you will need to be more selective and adaptive, for example with terms such as 'impact' or 'success'.
Operationalise concepts	This is how a concept might be measured. This could be a qualitative measure, or a quantitative measure.

The range of literature sources available

The good news is that there is a vast range of literature sources available. You will consider the quality of different resources in the next section on page 22.

Your literature search will probably be based around a variety of approaches:

- Obtaining relevant literature referenced in books and articles you have seen as part of your A-level Geography course.
- Scanning and browsing secondary literature in the Geography department, school library or local VLE.
- Visiting a local university library (access as a private researcher can be applied for in some libraries).
- General internet browsing, including Google Books and Google Scholar.

Books and magazines that you have easy access to are a good starting point, but bear in mind that they quickly age and sometimes lack the detail required. You will find that these often act as a trigger to access 'secondary' resources.

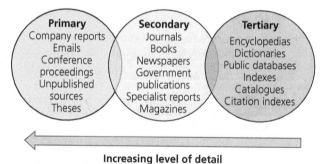

Figure 6 Available literature research.

Planning your literature research

It is important you think about the types and range of resources you might need at an early stage so that you can locate them. There are some examples in Figure 6. Often the literature search is very time consuming. Fortunately, time spent planning at this stage should be rewarded since this literature underpins many aspects of the investigation.

The first thing to do is to define the parameters (scope) of your research.

Table 8

Parameter of research	Examples
Broad topic area	Community change
	Glacial morphology
Focus area	Impacts of change
	Evidence of relic ice in the upland landscape
Geographical location	Communities within Cromer
	Nant Ffrancon
Literature type	Undergraduate books
	Geography Review magazine
Publication period	Last five years
	Last two years

You can then begin to work out the search terms that you could use in an internet or library search. Search terms can also be developed by using online dictionaries and encyclopedias to establish specific or technical language.

My progress

1 Based on your progress so far, complete the table below for your own NEA.

Identify concepts	
Define concepts	
Operationalise concepts	

2 You could use these concepts to create a concept or mind map showing the strength of relationships. Use something similar to what you did on page 13.

3 In the table below write down the parameters of your proposed research according to the headings listed. Use the information on the previous page to help you.

Parameter of research	My answer
Broad topic area	
Focus area	
Geographical location	
Literature type	
Publication period	

4 Now make a list of the specific search terms and phrases that you will be using in your literature search. Justify the selection of each one. Take your time with this, and if possible, work in small groups to develop some initial ideas.

Search term or key phrase	Justification/reason

3.2 Evaluating the reliability and provenance of different sources

We know that there is potentially a wide range of literature sources, but these can vary in quality. It's important to be critical when looking at sources.

Understanding provenance

It is a good idea to understand both *who* and *why* someone has produced a piece of research evidence. Start by considering the type of evidence – what is the source? Is it a published document, or an extract from a blog or forum? It could be a personal communication or an extract from a video posted online. No matter what type of source, for the fieldwork investigation it is best practice to be able to formally evaluate all documents and opinions.

Fake news? Questioning the reliability of evidence

Linked to the idea of provenance is recognition that different types of sources may have different amounts of bias. Generally, sources which are published in an official capacity are edited, and usually can be more relied on to be both neutral and factually accurate. You may need to use social media, for instance, as part of your investigation, and while it can give biased information it can also be very useful. With these sources, you need to ensure you assess their reliability and cross-reference them where possible (see Table 9).

Contextual knowledge

The other aspect of source evaluation is to consider what knowledge you already have that might support or challenge a view or another piece of evidence. For example, you might expect that deforestation in a small catchment increases flood risk, yet you may find that as part of your research there is evidence that this is not always the case.

You should never be worried that you find your results differ from what is written about a particular topic. In fact, it gives you more to talk about.

Examples

Below are four different sources of information on measuring and variations, in deprivation.

Source A: UK Government
The website lists various statistical releases and documents, including an infographic, which identifies how deprivation is calculated.

Source B: Consumer Data Research Centre
The CDRC was established by the UK Economic and Social Research Council as part of phase two of the Big Data Network. Data is available as a range of downloads for use with GIS, or ready-built maps. Raw data (numerical) is harder to source.

Source C: University of Liverpool
This article, in written prose only, is hosted by the University of Liverpool. It has several links to other studies, as well as a link to the study on which this summary is based.

Source D: Joseph Rowntree Foundation and Inspiring Local Change
An 84-page report exploring deprivation in selected UK cities written by a team of university authors. It contains a large number of maps.

Table 9 Sources and reliability.

Type	Source comments	Possible degree of bias
Academic papers	University authors, researchers. Peer reviewed and assessed. Generally technical and can be difficult to understand.	Likely to be least biased
Technical documents	Central government, NGOs, etc. High degree of credibility and wide range make these very good sources.	
Textbooks (including undergraduate level)	Large publishers, written by a team of authors and edited. Can be selective in the information and arguments presented.	Mixed
Magazines, newspapers and published leaflets	Wide variety – some very good, others more questionable. Will often reflect a political persuasion. Often up-to-date, but may not be easily accessible.	
Social media and personal communication	Twitter, Facebook, personal blogs and any 'unregulated' information on the internet. You need to be careful with some of these.	Likely to be more biased

Evaluating sources

Answer the questions below in the boxes for each source.

1 Look at the origins of the sources on page 22. How much do you trust the authors?

2 Which organisation has the data come from and should you trust it?

3 What are the advantages and disadvantages of each particular source?

4 How up-to-date do you think each one is likely to be?

5 What contextual information do you know that would be useful to be able to evaluate these sources?

Source A	Source B

Source C	Source D

My progress

When looking at sources of evidence for your coursework, you need to be able to quickly evaluate their reliability and provenance. We have started a draft checklist for you in the table below. Can you work out how to adapt, refine and improve these initial ideas? You could also rank them in a different order, based on how you see their importance to your own research.

Checklist – draft	Refined
Author and organisation provenance?	
Is the source clear and understandable?	
Is it neutral and unbiased?	
Is it up-to-date?	
Does it have a coherent style?	
Are the ideas referenced to other documents?	
Can I verify what is written?	

3.3 Literature review 1: using sources to find a model, idea or concept

Models are important as they allow you to test, validate and challenge a pre-existing assumption. They form a key strand in the purpose of the investigation and you will need to refer back to them again during your analysis and conclusions. You need to be open minded as to what a model is – they are often quite hidden within research sources and need some teasing out in order for them to be successfully utilised.

Model, idea or concept?

Models, theories, ideas and concepts are really just assumptions. They are someone else's thoughts that have been published (often using a supporting explanatory diagram and text) to say how something (or a set of processes) is expected to work or function. Often, they will form a testable set of ideas or assumptions that can be analysed using quantitative or qualitative evidence.

Being critical of assumptions

Published models, ideas and concepts may have a number of problems that any reader should be aware of. These can include the following:

- Ideas tend to be based on evidence from one geographical area, so results may not be directly transferrable to another location.
- Results may be modelled on a scale which is very different to the scale that you intend to study, so they are not directly comparable.
- Models can date quickly so the reasons behind an idea from ten years ago, may not be so valid now.
- Models and concepts, by their very nature, tend to simplify complex systems and processes that can be dynamic and complicated.

Examples

Finding these assumptions is actually easier than you think, once you have done some initial research. Figures 7 and 8 and Table 10 present different types of model that could be used as a testable idea or comparative context for a fieldwork investigation.

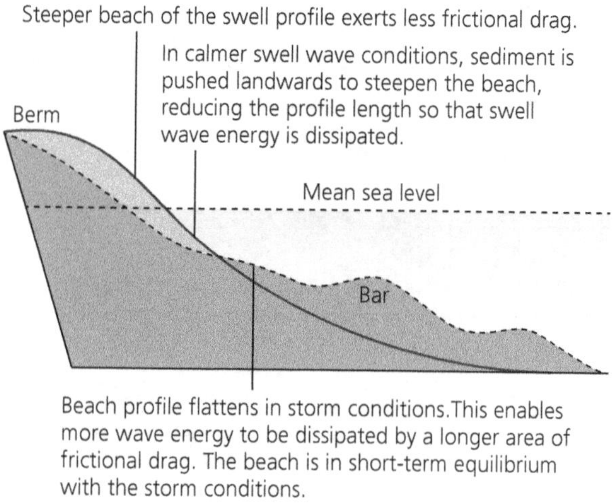

Steeper beach of the swell profile exerts less frictional drag.

In calmer swell wave conditions, sediment is pushed landwards to steepen the beach, reducing the profile length so that swell wave energy is dissipated.

Berm

Mean sea level

Bar

Beach profile flattens in storm conditions. This enables more wave energy to be dissipated by a longer area of frictional drag. The beach is in short-term equilibrium with the storm conditions.

Figure 7 A 'traditional', drawn model which includes a spatial element showing how a beach profile is expected to change between winter and summer, depending on wave type. There is a large number of this type of model for human geography as well, especially in specialist books.

Table 10 This is the 'Amazon effect model'. A table of data showing average changes in the increase and decrease of certain types of shops in 2017–18 on the high street. Quantitative data such as this can be 'tested' in other locations.

Risers	Net increase (%)
Barbers	624
Beauty salons	388
Tobacconists and vaping	381
Cafés and tea rooms	353
Nail salons	176
Restaurants and bars	174
Fallers	**Net increase (%)**
Pubs and inns	747
Banks	711
Travel agents	697
Post offices	577
Newsagents	364
Estate agents	349

The Hipster Index: Brighton Pips Portland to Global Top Spot

By Frederick O'Brien
Last updated on 19 Apr 2018
www.movehub.com/blog/the-hipster-index/

Figure 8 This is a much less obvious example of a model or concept, but dig into the online article and you can find a testable assumption. This kind of online blog can help you to find the original study or research and then find out where the concept comes from and then you can investigate further.

This urban model was developed in 1939 for US cities. It has now been used in an investigation *which compares the reasons for differences in urban quality of life.*

Look at the context of the model and write a few ideas around it about how useful you think such a model is going to be. In particular, think about the following: is it actually testable? How does it relate to quality of life? Is the model still relevant in modern western cities?

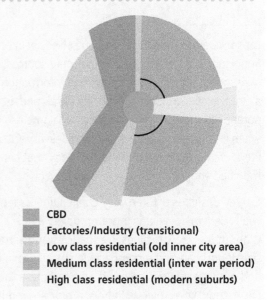

■ CBD
■ Factories/Industry (transitional)
■ Low class residential (old inner city area)
■ Medium class residential (inter war period)
■ High class residential (modern suburbs)

My progress

You are going to evaluate some possible models/concepts for your investigation. From your research so far, make a list of three or four models/concepts that might be suitable. Then for each model, rank them (best to worst) against each other and provide a justification as to why you might or might not use them. Think about their relevance in relation to the focus of your investigation.

Model/Concept/Assumption/Idea description	Rank	Justification for ranking decision and selection

Blaxter *et al.* (2001) suggests that, if there appears to be no research in your field: '... you should probably consider changing your topic. Ploughing a little-known furrow as a novice researcher is going to be very difficult, and you may find it difficult to get much support or help.'

3.4 Literature review 2: researching fieldwork design and 'reviewing the review'

Let's take a moment to remind ourselves again of the importance of the background literature sources. Not only should they give you contextual information on a place, as well as ideas for concepts and models, but potentially they will give you access to some useful ideas when it comes to survey design and delivery. Often the literature has a design and methodology that can be modified and adapted for your own project.

Survey design is often a very challenging aspect of the NEA (we cover this more on pages 31–37). Design is about location, timings, frequencies and sample size. If you can find a comparable study with a clear and manageable design use it as a template for your pilot study, then review and discard or amend it as you see fit. Some suggestions and strategies follow on the next few pages.

Examples of design

The key here is to read widely and get ideas from as many sources as you can realistically find. In this first example, a paper about perceptions of coastal management, the precise research methods are described. The authors base their design on previous work (see Figure 9); in Figure 10 they give details of the questionnaire itself.

In Table 11, the researcher is looking at how cafés are used as community spaces. This is an extract from their face-to-face interview which was intended to last approximately 60 minutes. Notice how the introduction includes ethical information and considerations.

Available online at www.sciencedirect.com

SCIENCE @ DIRECT•

Ocean &
Coastal
Management

ELSEVIER   Ocean & Coastal Management 46 (2003) 565–582

www.elsevier.com/locate/ocecoaman

Public perceptions and attitudes towards a forthcoming managed realignment scheme: Freiston Shore, Lincolnshire, UK

L.B. Myatt, M.D. Scrimshaw, J.N. Lester*

Faculty of Life Sciences, Department of Environmental Science and Technology, Environmental Processes and Water Technology Research Unit, Imperial College of Science, Technology and Medicine, London SW7 2BP, UK

Figure 9 Perceptions of coastal management paper – original research.

6. Research methods 6.1.

The survey instrument

A postal questionnaire entitled 'The Coast & You' was devised and standardised (for all three managed realignment case studies) for the purpose of collecting data in a format suitable for analysis. As detailed previously in Myatt *et al.* [3], the survey evolved from pilot work conducted at Brancaster West Marsh, North Norfolk [2] and influenced by a riverine environment and flooding attitude survey conducted by ARTICLE IN PRESS 570 L.B. Myatt *et al.* / Ocean & Coastal Management 46 (2003) 565–582 Middlesex University, Flood Hazard Research Centre [14]. In order to execute the survey at Freiston Shore, slight modifications of the original Brancaster survey were made in terms of amending the grammatical tense of some questions (i.e. from present to future).

Figure 10 Perceptions of coastal management paper – questionnaire details.

Table 11 Interview questions on the use of cafés as community spaces.

Objectives/ timing	Questions
Introduction (2 mins)	Nature of research and how it will be used • Research study aims to … • University research funded by the Centre for Business in Society at Coventry University. • Used in academic and industry publications, conference papers and a funders' report. • Recording for recollection purposes/quotes. • Anything said will be treated as confidential and anonymous so your personal data will not be passed on to anyone else. • I'll nod a lot because I want to hear you, not me. • Think of this as an informal chat. I'm interested in your own reflections on your career. We want honest views and opinions about the topic. • So, before we start, please fill in the consent form.
Café history (10 mins)	Can you tell me a little bit about how this café came about? • When was it opened? • Part of a chain? How much contact do you have with other branches? • How did you end up getting involved? • What did you want to achieve by setting up this café?

Customising, modifying and making it manageable

It may be that the design or methodology you have found is too big, too complex or is simply unworkable for the scope of your NEA, in which case you can adapt and simplify it to fit your needs. In Table 12, the student has found a walkability audit through their research and has then simplified it (see Table 13) to make it more manageable for their proposed fieldwork.

Table 12 Original walkability audit.

Measuring urban design qualities scoring sheet	Auditor		
Street From	Date & time		
Step	Recorded value	Multiplier	(multiplier) × (recorded value)
Imageability			
1. Number of courtyards, plazas and parks (both sides, within study area)		0.41	
2. Number of major landscape features (both sides, beyond study area)		0.72	
3. Proportion of historic building frontage (both sides, within study area)		0.97	
4. Number of buildings with identifiers (both sides, within study area)		0.11	
5. Number of buildings with non-rectangular shapes (both sides, within study area)		0.08	
6. Presence of outdoor dining (your side, within study area)		0.64	
7. Number of people (your side, within study area)			
Walk through 1			
Walk through 2			
Walk through 3			
Walk through 4			
Total			
Total divided by 4		0.02	
8. Noise level (both sides, within study areas)			
Walk through 1			
Walk through 2			
Walk through 3			
Walk through 4			
Total			
Total divided by 4		−0.18	
Add constant		+2.44	
Imageability score			

Table 13 Simplified walkability audit.

Street name Grid reference Note – record on both sides of street	Date and time		
Imageability identifier	Scored value	Weighting	Total score
Number of outdoor seats, courtyards and green spaces		0.4	
Number of historic buildings		1.0	
Number of outdoor dining places (coffee shops, restaurants)		0.6	
Number of buildings with modern architecture		1.0	
Pedestrian count (average of 5 counts, 5 minutes)			
Noise score (average of 5 readings at intervals)			
Geolocated photos – completed?			

Complete the activity as part of your planning and review of published sources of design and methodology, relevant to your topic. Two examples have been included for you.

Source/Resource – design and method	Modification required? Quality?
Guardian article about gender and attitude towards recycling from October 2018: www.theguardian.com/environment/2018/oct/05/real-men-dont-recycle-how-sexist-stereotypes-are-killing-the-planet Followed links to original papers to get ideas for the method.	Decided to email the main author to see if I could get a copy of the original survey design to find out how perception was measured. Study was based on new research so that gave confidence in the quality.
Stratford-Upon-Avon visitor survey report: www.stratford.gov.uk/doc/206624/name/Stratford%20upon%20Avon%20Visitor%20Survey%20Final%20Report%202015.pdf Includes details about both sampling and the survey itself.	Was able to use the information in the report. Simplified approach to concentrate on visitor profile, so I can use a modified approach in my report. Stratford is very different to my study area, so I have concerns about direct comparability of the methodology.

3.5 Using other (experimental) data from research and literature sources

The literature survey may also turn up other experimental data that can be used as an important source to test against your own information and ideas.

Why would you want to use such data? Well, the answer is simple – it allows a comparative context for you to test whether your investigation produces similar or different results. You can then write about the possible reasons for the degree of similarity or difference.

Data everywhere

There are lots of examples of different data and information in a published format. In many respects you need to think about what type of data and information you might want: visual; graphical; textual; numerical? Each different type may have its own separate advantages and disadvantages.

Figure 11 shows an extract from an online Tourism Business and Leisure Survey (2018). In it a range of different tourism businesses were asked if their use of social media has increased (blue) or decreased (pink), between 2017 and 2018. This could be used in a comparative context to see if your own, locally collected data, matches the trends shown here.

Figure 12 shows published data from an academic journal investigating public attitudes towards managed realignment. The study was based in Lincolnshire and was undertaken more than 15 years ago. This data could be directly compared with another coastal location, assuming that the same types of questions and attitude statements were used. It could also be used to see if the public's knowledge or perceptions of management realignment have changed since the original survey had been undertaken.

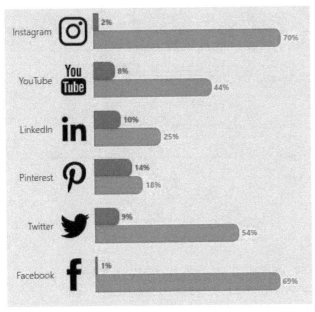

Figure 11 Extract from an online Tourism Business and Leisure Survey (2018).

Managed realignment attitude statements

Statement	Agree (%)	Disagree (%)	Neither agree nor disagree (%)	Non-response (%)
The saltmarsh will absorb wave energy from the sea	38	8	44	10
Managed realignment is only a short-term coastal defence option (not the view of the Environment Agency)	34	11	43	12
The managed realignment scheme will not reduce the flood risk in this area	23	35	32	10
The managed realignment scheme will provide more protection from coastal flooding than a sea wall	37	21	35	7
Keep the present line of defence, it is equally effective as the managed realignment scheme	30	22	37	11

Figure 12 Public attitudes towards managed realignment.

Revisiting the proposal form

Having got this far in the planning stages, it is a good idea to revisit your proposal form and update it based on the information you have found in relation to the research components. Use the framework below and then amend your form accordingly.

1 Do you need to modify the overall title so that you have more secure links to the specification content?

______

______

______

2 Is the planned investigation hypothesis or question(s) still valid, especially if you need to link it to some other research data (for comparison)?

______

______

______

3 Is your planned methodology still going to work or do you need to tweak it in order to align more closely with the **empirical research data**?

______

______

______

4 If you are collecting data as a group or with someone else, will their data (which you should treat as secondary data) be useful as part of the comparison?

______

______

______

My progress

Identify and select two or three other experimental data sources that might be relevant for your enquiry. Write down some advantages and disadvantages of each.

Source 1	Advantages and disadvantages
Source 2	**Advantages and disadvantages**
Source 3	**Advantages and disadvantages**

Section 4 Fieldwork planning and design

4.1 Choosing a design: sampling, sample size and a management scale

This is the part of the fieldwork investigation where you need to think carefully about what, how, when and where primary data can be collected. Remember, primary data is first-hand and individual to you, the investigator.

There is a difference between design and method. Design includes consideration of location, spacing, timing, and so on, whereas method is more to do with the actual equipment or recording sheet used.

Why sample?

Good sampling is integral to good research design. Sampling is nearly always needed since you can't measure everything. Sampling is an accepted shortcut – so a sample is a selection of data from a larger population of things, items or people than you could possibly measure. Examples of populations are: survey points within a drainage basin; people in part of a town; or the stones found in an area of beach.

Figure 13 In a busy urban environment like this there simply isn't enough time or resources to collect data about all the processes that might be going on.

Sample size

It is always a good thing to be able to say why you took a certain number of measurements. Why, for instance, might you count things in 20 rather than 30 quadrats, or why might you undertake 15 rather than 20 environmental quality surveys? Yet the question: 'How large does my sample have to be?', is surprisingly difficult to answer and requires thought since sample size can affect both the strength and reliability of conclusions. In fact, it is often difficult to determine a sample size without a pilot survey (see page 34).

Environments with a large variability (such as lots of different sizes of stones or wide differences in pedestrian flows) need larger sample sizes. Whereas in locations where you find much more similarity, you can be more secure in the fact that you will need a smaller sample size.

There are several online tools that can help you with sample size calculation (see Figure 14 for an example). Although designed for questionnaires, they are also useful for other environments, although you will need to approximate the population. The basic rule to remember is that more samples generally mean greater reliability.

Sample Size Calculator

Qualtrics offers a sample-size calculator that can help you determine your ideal sample size in seconds. Just put in the confidence level, population size, margin of error, and the perfect sample size is calculated for you.

Confidence Level:
95% ▼

Population Size:
10000

Margin of Error:
10% ▼

Ideal Sample Size:
96

Figure 14 An online sample size calculator.

You will also need to decide an appropriate area to sample within (this is often called the sampling frame).

Using the running mean approach

The running mean is a simple technique that allows you to judge whether or not you have enough measurements or counts. The running mean, or moving average, is a statistical approach which takes into account the natural variation that might occur at each site or sample point. In Figure 15 the moving average (running mean) is shown by the orange line. It is calculated by finding the mean of the first two readings, then the mean of the first three readings, then the mean of the first four readings and so on. The mean values will fluctuate each time, but will gradually settle within a closer limit, until the point is reached where adding to the sample only has a very small effect on the mean. You can assume at this point that the number of repeats is adequate (in the case below, shown by the dotted black line at 16 samples).

Different types of sampling

Sampling is about trying to make sure that the measurements taken are representative of the entire population (which is often made up of a large number of items, people or factors). Sampling strategies try to reduce the risk of bias, which introduces uncertainty and unreliability. There is a lot of written information about sampling, so this workbook will only give a brief overview. The box here gives information about the three main sampling types. Remember, many NEAs don't give this aspect enough consideration. The sampling framework must be very closely aligned with the geographical focus being investigated. Frequency and timing of observations must also be given due consideration so that the work is manageable.

There are more considerations of sampling in Section 5.5 Evaluating design and fieldwork methodologies (page 47).

Most commonly used types of sampling

Systematic sampling
Sampling at regular intervals, with points evenly distributed. You would usually use this when you expect a change from one area to another, such as along a road.

Stratified sampling
This is used when you think there are subsets or groups within a population that are different to each other. You base your sample design around the proportions of those groups to reduce any bias.

Random sampling
Using a random number or creating a grid so that any place, or something/someone in a population has an equal chance of being selected. This approach is used if nothing is known about the area being studied.

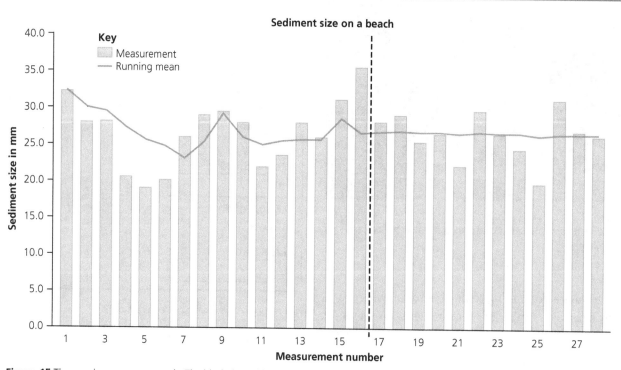

Figure 15 The running mean approach. The black dotted line indicates a more stable running mean so enough samples have been taken.

Now that you have considered your sampling strategy, revisit your proposal form. Remember, it's a document which should be seen as a planning tool and very much a work in progress.

Have a look at this part of the form again. Can you make any changes to the section where you have been asked to write about your 'planned sampling strategy'?

> Planned methodology – indication of qualitative and/or quantitative techniques including primary and, if relevant, secondary data collection techniques. indication of the planned sampling strategy or strategies.

My progress

Complete the comments section in the table below to show that you have given consideration to the following ideas in relation to sample size. Copy the table onto a sheet of A4 paper in landscape format if you need more space.

Sample size considerations	My responses
How much time do I want to spend collecting data in this location?	
How much variability do I expect in this population or environment? (And how might I find this out?)	
Will I be collecting the data as part of a group or individually? If I am part of a group how do I show individuality and my own contribution?	
Is there additional secondary data that can be used to support my analysis?	

4.2 Recommending the pilot survey

A pilot survey is a small-scale feasibility 'test' of your design, equipment and methodology. Pilot surveys are often associated with human geography (especially questionnaires and surveys), but are valid for any type of primary data that might be collected during the course of an NEA.

Conducting a pilot survey prior to the actual, large-scale survey provides many opportunities for the researcher. The most significant of these is the exploration of the particular issues that may potentially decrease the reliability of the data collection. In questionnaire design for instance, these include the sequencing and format of questions.

Uncovering the need for a pilot survey

A pilot survey should give a clear insight into the feasibility and timings of your investigation, as well as offering a practical opportunity to trial. Have a look at how a pilot survey could help in the following example (Figure 16).

Reflecting on your pilot survey opportunities

A pilot survey really does come recommended. How would using a pilot survey help you to overcome some of the following challenges?

Example	Relevance to my investigation
Refine recording sheets	
Identify logistical/access problems	
Ethical considerations	
Skills needed to use particular equipment	
Familiarity/navigation around the survey area	
Collecting enough data within the time allowed	

Firstly, reconnoitre the area, undertake assessment that it is fit for purpose – including risks/ethical considerations

Find out when there are most visitors (days and times)

Speak to the café owner and find out whether they can be interviewed at a convenient time

Pre-test questionnaires

Find out whether you can contact the managers of the forest park (Forestry Commission)

Collect any leaflets and look for links to online supporting resources

Try to work out the best places (sites) to collect data

Figure 16 A study to examine visitor profiles to a woodland park.

Take some photographs in case the weather is poor at a later date

Complete a brief risk assessment and consider ethical issues

My progress

Use this space to think about your pilot survey, and to make some comments about how you adapted your approaches and sample design based on how your pilot survey worked out.

4.3 Quantitative and qualitative fieldwork data and information

Different types of data?

Data comes in many different forms, including primary and secondary. In the context of the fieldwork investigation you need to consider primary data as first-hand – that is, data that you have collected yourself, or as part of a group (see page 36). Your project *must* include a proportion of primary data. It cannot all be based on secondary data.

Knowing how many different types of data to include is not straightforward. Relying solely on one primary source of evidence is, at least in terms of an experiment, risky, since any conclusions will just be based on those results. At the other end of the spectrum, too many separate techniques are also problematic as they will be unlikely to have enough depth. As a guide, between two and four, plus some secondary data, would be a reasonable number.

Quantitative data

Quantitative data consists of numerical and frequency data which can be tabulated, tested statistically or converted into charts and graphs. Many investigations will make use of this type of data, either in a primary or secondary form.

Quantitative data has several advantages:

- It is often more straightforward to analyse using traditional statistical techniques.
- The range of graphical and cartographical techniques is more 'off the shelf' and there is a lot of easily accessible information.
- Analysis can be quicker and more automated, for example using a spreadsheet or algorithm to process data.
- Larger samples make conclusions more robust.
- It is good for situations where standardised comparisons are needed.

Qualitative data

Qualitative data is non-numerical observations and descriptions of places and geographic processes. It includes in-depth interviews with people (respondents), photographs, text extracts (from poems, novels, travel guides and so on), paintings and even music which can evoke a meaning about place.

Qualitative research is the method of choice when the research question requires an understanding of processes, events and relationships in the context of the social and cultural situation. Qualitative research often gathers data from relatively small numbers of people or places, and can provide a 'micro' view of whatever issues you are examining.

Figure 18 Pictures are often an underused element of projects. Think carefully about the ethical considerations of taking pictures of people in public spaces, and private areas, such as shopping centres. But pictures including people really help to tell a story. In the physical environment, work on trying to show scale and processes.

Look at the images in Figure 18. What is their 'story'? Why has the photographer chosen to show certain aspects? What is not shown in these photos?

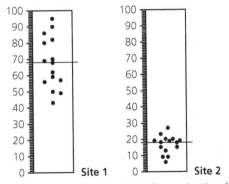

Figure 17 Plotted quantitative sediment size data from two different stretches of the same beach. Note results are in mm.

There are a variety of ways that imagery as qualitative data can be used as part of an NEA. Here are some examples.

1 Recording alternative views of a place (what the tourist brochures do not show you), either in a town or city. In urban environments this can be particularly revealing since some buildings or areas are deliberately excluded from promotional material.

2 Looking into the use of heritage marketing or identity marketing and how this type of approach is used to suggest a 'great day out'. Visiting a local tourism information centre can reveal a great deal about a place and its identity, as well as giving you access to brochures and flyers (that include images) for events in the area.

3 A survey of local products and produce can document how image (for example, 'localness', 'naturalness', 'wholesomeness' or regional speciality – 'Balti county', and so on) is used to promote place identity through food.

4 Exploring the 'tourist gaze' by observing how visitors to a place interact with its features and characteristics. Then record areas and buildings that are most photographed as they are perceived as iconic, characteristic and reflect the identity of a place.

Qualitative data also has significant advantages. It has richness and detail, which is especially useful for allowing respondents to explain more fully what they mean. Places and processes can also be seen more in their proper geographical context, which allows for a more holistic understanding. Qualitative data and evidence is also very useful for situations when detailed understanding is required – so this is especially relevant to places.

Making group data 'personal'

Sometimes fieldwork data is collected by a small group as part of a field trip. This approach is often used, but it is good practice to show personalisation and individuality of any group data and group data collection methods. There are several ways that you might do this.

- You could try to show evidence of your own data design aspects (for example, colour-coding your contribution in a group data set).
- You could show your own personalised maps of places, rather than using 'group-provided' versions.

- You could show personalised sampling strategies and investigation design elements (perhaps even based on your own pilot study).
- You should always adapt recording sheets, rather than directly using the same ones provided by your teacher or a field centre.

It is also good practice to show that you have personalised your literature research. If you are going to a field centre, for instance, then it is a good idea to make sure that you have completed aspects of the literature survey before your arrival. The strongest evidence is often when you are able to combine what you have read, with what you actually find as a part of a field trip.

Top tips for the collection of qualitative and quantitative fieldwork data

Careful fieldwork planning is essential for a successful outcome. Below are some suggestions to help you when you are planning your fieldwork approaches, to collect either quantitative or qualitative data.

- *Does it answer your question?* A pilot survey should confirm that data collection is appropriate, realistic and manageable for your intended focus. You need to be able to justify why you have used particular approaches.
- *Depth versus range?* There is a trade-off between how many different things are measured versus the number of separate measurements (i.e. repeats). More repeats improve the reliability of your outcomes.
- *Amount of data.* Remember that data needs to be processed and then analysed. The volume of data affects the value of that analysis.
- *Spatial coverage.* More samples in more areas or locations, improves your spatial coverage leaving less gaps. However, there is a trade-off between coverage and frequency of observations.
- *Timing of surveys.* Some processes change quickly, and your data collection should ideally try to capture those changes. Otherwise you should use secondary data to show you have thought about the changes.
- *Ethical considerations.* These may impact on the location and timing of observations, as well as the ways in which you interact with people in the built environment.
- *Risk assessment.* Although there are no marks directly for completing a risk assessment, it is required and should form part of your methodology. Often tables are used to summarise the risks and their management.

Qualitative versus quantitative and primary versus secondary

It can be quite difficult to separate out what different types of data actually are, and how they might be classified. Have a think about the following data types which are not easily categorised. Do some research and come to your own conclusions. Complete the table where there is space. An example has been completed for you.

Example data type	Classified?	My notes
Census	Secondary, quantitative	It is secondary when it has been processed and published, but some researchers say that when it is tabulated ('raw') then it is primary. Tricky!
Social media, e.g. Twitter		
Someone else's photographs		
An interview you found on YouTube		
A newspaper report when an event has been witnessed		

My progress

Complete the relevant sections of the following table so that you have an audit of the different types of data that you intend to use in your individual investigation. Make a note as to whether they will be primary or secondary.

	Qualitative	Quantitative
1		
2		
3		
4		
5		

Section 5 Writing up your geography fieldwork investigation

5.1 Introduction to writing up

The need to maintain focus

As you have already seen from the other sections of this book, the majority of skills that you need to succeed in this fieldwork investigation have already been developed in your AQA Geography course. The introduction made clear (see page 7) that most of the available marks for this Unit (54 marks out of 60) are for AO3. The analytical skills you need are also essential for a good mark for Paper 3.

AO3

Use a variety of relevant quantitative, qualitative and fieldwork skills to:

- investigate geographical questions and issues
- interpret, analyse and evaluate data and evidence
- construct arguments and draw conclusions.

However, this is a much longer piece of work than Paper 3 and you will need to make sure that you maintain focus on what you originally set out to do. In such an extended piece of work, there is sometimes a tendency for students to drift and lose focus. Remember that what you have planned to do has been signed-off by your teacher and will be submitted if a student's investigation is part of the sample required for moderation.

As the NEA will be much longer than an exam answer, you will also have the opportunity to consider the different ways in which geographical concepts, processes, ideas and theories can be explained and interpreted. You may, as a geographer, come to the conclusion that some issues are more important than others in explaining an aspect of your investigation and you will need to weigh up, or evaluate, those different issues. Your literature research and review will have ensured that you looked at different ideas, and you will have thought about the evidence there is to support or challenge each one, as we saw in Section 3. You will have had plenty of time to think about the factors which may impact on your findings and to find plenty of relevant, accurate and detailed evidence to support judgements. There is no one, single book that will provide you with a readymade answer: there is no shortcut to this. You have to carry out the research so that you are in a position to begin writing up your work.

As you begin to write always remember your focus – that is, what you set out to do at the start of your fieldwork investigation. This should also ensure that you write relevantly for the 3000–4000 words that are recommended by AQA (although the majority of students, rightly or wrongly, exceed this recommendation).

Some advice:

- You will need to ensure that there are no parts of the write-up where you lose focus on the actual question. There should not be any sections where there is just description, or processes and concepts that are not related to the purpose.
- You need to ensure that you have considered different aspects and interpretations of what you have found, and have not produced just one single argument.
- All of your arguments will need to be backed up by detailed and accurate evidence and you will need to apply your knowledge and understanding when relating your findings to concepts and theories, and the wide context of your investigation.

When you are ready to begin writing up you need to have available:

- all your reading and literature research
- all your fieldwork evidence, with individual and group contributions separated, if relevant
- all your secondary data and evidence. This might be used to support and challenge ideas
- additional information you want to include such as sketches, handwritten notes, logs, diaries and photographs.

Creating a simple plan and schedule

How will you bring all this together? The best way to do this is to ensure that you have a clear plan and detailed schedule with targets.

Balancing writing your investigation with other school or college work and social commitments can be difficult. Building a plan or schedule helps you manage your time more effectively. It gives you a clear idea of how much time different sections should take, and therefore helps you to organise the delivery of the NEA around your life.

A useful starting point for your plan might be to construct a spider diagram or mind map which outlines the themes and ideas you are going to cover, as this will give you the chance to check that you have considered all the issues you need and that they are all relevant to your actual question. This could also show connections between geographical processes as you expect them to occur in your given environment – how the geography 'works'.

Let us assume that you had chosen to investigate the factors which influence the attitudes of a community towards gentrification and changing place identity. Your spider diagram would provide you with a list of factors that may influence attitudes, both local and perhaps national. However, it should also show any links between factors and it would be helpful, in light of your research (and/or pilot study) to show the relative importance or to weigh up the factors so that you have actually assessed the reasons and not just made a list. The result might look something like Figure 19.

You could then indicate the importance of each factor with a number and provide a brief justification as to why you consider it to be of greater or lesser importance.

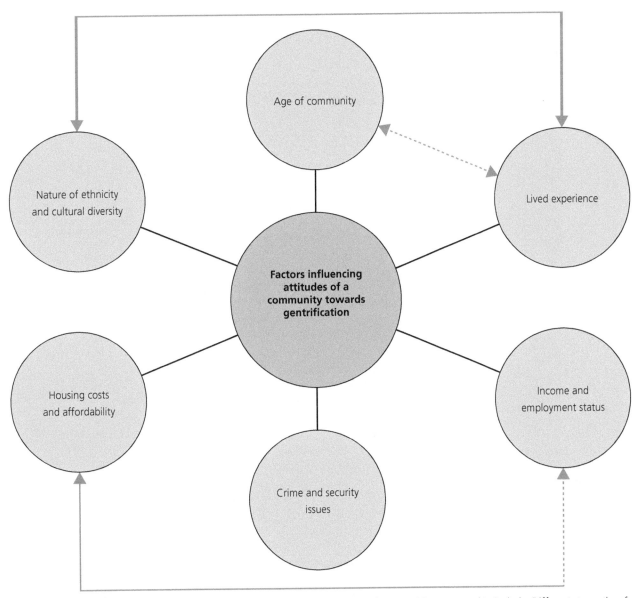

Figure 19 Spider diagrams can help you sort your ideas and make sure you consider everything you need to include. Different strengths of links can be shown by dotted lines or thinkers arrows for instance.

Spider diagram

Construct a spider diagram for your own title showing the geography that you think 'binds' your investigation.

- What are the main issues?

- What links are there between issues?

- How important is each of the issues in answering the question? Put a number by each factor and explain its importance.

Now go on to develop a schedule – see page 78. An effective schedule means that you have control of the dates by which you will complete each stage of the investigation.

5.2 Developing an appropriate writing style: avoiding narrative and description

If you look at the mark scheme in Section 1 you will see that write-ups that are largely descriptive or narrative do not gain many marks. As for most pieces of technical writing, it must have a persuasive argument!

The plan you have drawn up should help to stop you just simply telling a story, but it is important that you are clear as to what is meant by 'narrative' and 'descriptive' before you start writing.

Look at paragraphs A and B that were written as part of two different students' fieldwork investigations. They are both extracts from the final parts of their coursework.

Which of the two paragraphs has an analytical and technical feel, and which is more descriptive and has sweeping generalisations?

Paragraph A

Overall the impacts of the redevelopment are highly beneficial to the economy of the local area. There has been a rapid increase in the number of people living in the area over the past few years. This has led to a positive multiplier effect, which has the effect in economic terms of increasing local peoples' spending power…

After reading articles looking at the impacts of regeneration, I'm sure that my results show all positive effects. Some of what I have researched says that effects might be negative, but my data shows positive correlations between how people feel and the economic impacts…

Paragraph B

It can be argued that the study areas in west Norwich are attractive places considering the evidence in the photographs which clearly demonstrates high amenity value, especially in terms of green spaces. This primary evidence strongly supports the 'shared community activities' component of the Egan wheel, and questionnaire data (Figure 3.2) supports the conclusion that people utilise this space frequently (modal usage = 3 × visits/month)…

To summarise, west Norwich has attractive features, but these are not used or shared equally by all members of the community and therefore the strength of this conclusion must be tentative only. In addition, when the time aspect is evaluated, then the 'space' changes during the course of a day. Night-time, according to the primary and secondary evidence is far less attractive…

Knowing the 'rules' and using connectives

The fieldwork investigation is an exercise in geographical communication. Generally, you will write well if you:

- know why you are writing
- know who your audience is
- model your approach on another piece of similar technical work
- follow a logical and manageable structure
- overall, feel confident.

One key strategy you might use to improve the quality of your writing (and its style) is to think about connectives.

Table 14 Using connectives strengthens the quality of an argument, consequently connectives are key in persuasive writing see Paragraph B highlighting.

Balancing connectives	Evidence connectives	Cause and effect connectives
alternatively although however instead of nonetheless though whereas while	for instance in the case of revealed by such as	as a consequence as a result of consequently inevitably resulting in the effect of this is this results in this, in turn, causes

Use of the first person: the 'I'

Contrary to popular belief, fashions change in academic circles. It has become acceptable again to use the first person (the 'I') in essay and coursework writing, but it still divides people. Using the first person can bring a very personal dimension to the work which is really beneficial.

Others would strongly argue that you should stick with the third person in your investigation (the inclusive pronoun 'we' is always at your disposal, don't forget). It does, in most instances, read more technically correct.

The choice is yours, but have a look at the following example sentences and see which one you prefer.

My results matched my hypotheses so I am able to conclude that pebble size does vary significantly between X and Y.

The evidence supports the conclusions, therefore it is possible to state that pebble size does vary significantly between X and Y.

Reviewing phraseology and sense

Read and review the following generic sentences that could be used in the various stages of an NEA. In your review, think about:

1 Which of the sentences has clear sense and correct grammar?

2 Which sentences are more analytical (rather than descriptive)?

3 Which sentences use colloquial (informal) terminology? This is not really appropriate at this level.

1 X plays an important role in the maintenance of ...

2 Smith (2005: 227) shows how, in the past, research into X was mainly concerned with ...

3 Crime is becoming increasingly prevalent in society ...

4 The theory suggests that the hypothesis is correct because of the crime rate so theoretically it rises when an event is on ...

5 Areas with a low deprivation score will be most in need of regeneration ...

6 A considerable amount of literature has been published on X. These studies ...

7 This project is going to look at the coastal processes operating in north Norfolk ...

8 One possible implication of this is that ...

9 The results of the questionnaire definitely prove my hypothesis ...

10 The town lies at the south of the River Ouse and has some boats which go to France, for people who want to have a sunny holiday ...

My progress

Having completed the activity above, now apply this to your own focus.

Title/Aims/Hypothesis (of my fieldwork investigation):

_____

In the space below, simply practise writing a few sentences which include technical language, connectives and good grammar. They must also make sense geographically!

_____
_____
_____
_____
_____
_____
_____
_____
_____
_____

5.3 Writing the introduction and preliminary research

We have already looked at the specific requirements of the AQA Area 1: Introduction and preliminary research in Section 1 when we reviewed the marking criteria (see page 8). But it is also useful to note that there are many ways to introduce an academic piece of work and academic writers do one or more of the following in their introductions:

- Establish the context, background and/or importance of the topic.
- Indicate an issue, problem or controversy in the field of study.
- Define the topic or key terms.
- State the purpose of the essay/writing.
- Provide an overview of the coverage and/or structure of the writing.

For your fieldwork investigation, the introduction will likely contain one or more of the following additional components, as per the mark scheme criteria (page 8):

1 Clearly stating the research question(s), aim(s) or hypothesis(s).

2 Explaining reasons for the writer's personal interest in the topic and explaining the significance and value of the proposed study.

3 Giving a summary of the relevant literature.

4 Defining certain key terms.

Striking the right balance: place versus theory

Some investigations get bogged down at the start with too much description of a place rather than a set of geographical processes that are going to be investigated in a location. You need to think carefully about the balance here. Use the proportions in Figure 20 as a guide.

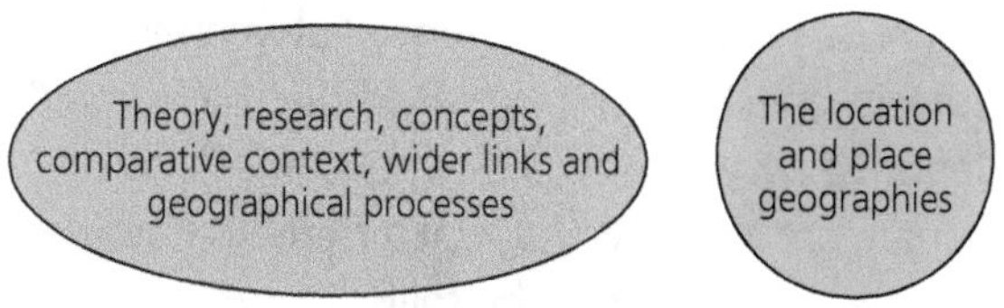

Below are a few examples for you to think about.

Table 15 Concepts linked to fieldwork investigation topics.

Figure 20 More of your introduction should focus on the ideas around the geography: processes, concepts, theories, and so on, rather than the place itself. This should help you to avoid problems with too much description at the start.

Using the specialised concepts and seeing the bigger picture

It is a good idea to introduce at the start of your investigation, a *geographical* reason for undertaking your study. Remember that in the first section there is a requirement for you to introduce the broader geographical context.

Specialised concepts are a useful way to refer your investigation to a broader geographical context. The specialised concepts are synoptic. They are the language that is used by professional geographers and academics. The inclusion of specialised concepts linked to fieldwork *could* add an extra dimension or layer of sophistication.

Your purpose should also consider the broader geographical context. This is really the importance of why you chose to study something – in other words, why are you actually bothering to carry out your investigation? If you have been counting stones on the beach, what's the point? See the activity on page 44.

gain understanding of specialised concepts relevant to the core and non-core content. These must include the concepts of causality, systems, equilibrium, feedback, inequality, representation, identity, globalisation, interdependence, mitigation and adaption, sustainability, risk, resilience and thresholds

Figure 21 Specialised concepts in the AQA GCE Geography specification.

Concept	Meaning	Examples of fieldwork investigation topics and ideas
Resilience	The capacity of a system to experience shocks, while maintaining the same function, structure, feedbacks and identity.	<ul><li>Flood risk and resilience</li><li>A local high street and resilience</li><li>Sand dune ecosystem biodiversity and coastal development</li></ul>
System	Understanding how components which make up a system influence one another within a complete entity, or larger system (boundary).	<ul><li>Inputs, outputs and stores within a very small catchment</li><li>Economy of an urban centre as a system of inputs and outputs</li><li>Understanding carbon flows in a woodland ecosystem</li></ul>

A focus reminder

In Section 1 we reviewed the title selection procedure as well as the difference between a hypothesis, question, and aim and how they should be identified by you. This is an important decision since there are subtle yet crucial differences between them. Have a think about the following:

- An 'issues-based' NEA inevitably considers a complex idea or argument. Does the title take into account this difficulty and allow for sufficient depth of argument?
- A hypothesis may not be suitable in every situation. It may only be a good idea to use a hypothesis where there is appropriate numerical data. A hypothesis can also be used as a tool for part of the analysis rather than the overall focus of the introduction.
- Should questions be subdivided? If they are, the focus can become more manageable, but it runs the risk of becoming too wide and unfocused. Studying too many variables can make it harder to make clear conclusions.
- Is there a maximum or minimum number of words for a good title/aim/question/ hypothesis?

Identifying the broader geographical context

Have a look at the table below which includes some topic ideas. Try to identify the possible wider geographical context as to why these might be studied. The first two have been completed for you. Note that there are many different reasons that you could include here.

Topic	Wider geographical context
An investigation into flooding risk in X	Climate change will increase the likelihood of more storm events in some places so it will change the flood risk probability.
An investigation into the success of regeneration on high street Y	High streets are under pressure from online retailing in particular. For places to be resilient in the future, rebranding must encourage people to have an 'experience'.
A study into the variations of quality of life between area A and area B	
Investigating the evidence of relic glaciation in area T	
Success and failure on the high street: where next for its identity?	

My progress

Make a list of key elements to include in your fieldwork investigation introduction. These will likely form sub-headings within your project. You need to be able to justify each one, as follows:

Sub-heading	This is relevant to my project because:

5.4 Demonstrating appropriate methods of field investigation

Showcasing good practice

You have already explored Fieldwork planning and design in Section 4 (pages 31–37), so you should be familiar with the terminology. At this stage, when you are attempting your writing up, you should have done most, if not all of your primary data collection.

Here is a quick reminder of ideas that you are targeting from the top band of the AQA mark scheme:

- detailed use of appropriate methodologies
- thorough and well-reasoned data collection approaches, relevant to the aims of the investigation
- consideration of timing and frequency of observations for good quality data and information
- reliable data collection methods (repeat sampling for reliability and/or accuracy).

All of these elements need to be ticked off in order for the highest marks in the top band to be achieved. Have them in your mind as you produce your write-up and try to provide clear references to the ideas and use geographical terminology. Adapted, they may even provide useful section headings.

The pros and cons of using tables

Fieldwork investigations often use tables to showcase a methodology and data collection. The key thing to get right with a table is the column headings. They need to be well considered and could be related to the criteria in the mark scheme or the variables under investigation. Importantly, you need to set up the table so that each method is clearly identified and there is also an opportunity for giving a reason, for example how or why these methods help answer your question.

The alternative approach to the use of a table is to use continuous prose. This frees you from some of the structural constraints of a table, but may be harder to deliver. A danger is that you might become too wordy, lack focus and have an incoherent structure.

Photographing equipment and measurements

There are several reasons why you should consider using photos in your methodology:

- They add individuality and personalise your approach to the write-up.
- They can show how you used a particular piece of technical equipment which might be easier than explaining it in words.
- They may demonstrate sampling, or a particular approach taken to ensure accuracy.

- They let the reader see that you have actually done primary data collection – so they provide evidence for the measurement(s) and data collected.

Some examples of how images and photographs might be used are found in Figures 22–23.

Figure 22 The practice use of a piece of equipment, for example showing how measurements were taken, is useful. It shows attention to detail and validity.

Figure 23 Here, a more unusual colorimetric method is being used to show dissolved organic carbon in stream water. It may, for example, show the difficulty of colour-matching.

Evaluation, risk assessment and ethical considerations

Many students include aspects of evaluation as part of their methods table, but you need to think about what type of problems and limitations are relevant and appropriate at this stage. Avoid falling into the trap where you say things like 'if I had had more time I would have done …' or, 'because it was raining there were not many people on the beach …'. Evaluations need to be more sophisticated, for example related to subjectivity of making semi-quantitative decisions. You will cover this more fully in Sections 5.9 and 5.10 as well as in the next Section, 5.5.

You should always do a risk assessment and include it as part of your work. It doesn't have to be very long, perhaps a maximum of a third or half a page, and it can be presented as a table. It doesn't carry any marks directly but it is *always best practice* to include one.

Ethical considerations appear in Area 4 of the AQA mark scheme so you may choose to include them in a later section, but they are also relevant at this stage.

Take a look at the list of technical terminology that might form part of your methodology and data collection write-up. Using a line, match the word to the correct definition. You can confirm the correct answers by using the glossary at the end of this book.

Reliability   An inclination or prejudice towards or against a specific finding or outcome.

Stratified   The number of measurements and may include considerations of timings.

Qualitative   A practice or trial survey to ensure that data collected will be appropriate.

Pilot survey   Longer open-ended style of questioning someone (open questions).

Bias   Information which is subjective or does not have any number(s).

Accuracy   The consistency or reproducibility of a measurement.

Frequency   The closeness of measurements of a quantity to that quantity's actual (true/real) value.

Interview   When you take into account known information about the characteristics of a sub-population.

This is not an exhaustive list. Think carefully about using the correct technical language to explain and justify *your own* methodologies and data collection.

My progress

An important, and personal, decision to make here is whether you use a table or continuous prose. There is not a single or simple answer, instead you need to weigh-up the various options and make the best decision given your own circumstances. Evaluate the options and then circle which one you think is the winning option for you.

Table	Continuous prose (text)
Advantages	Advantages
Disadvantages	Disadvantages
Winner?	Winner?

In this part of your write-up you need to think about the data collection methods themselves (let's call those 'methods', like using a sound-meter) and also the design aspects of data collection which include ideas around sampling, timing of observations, and so on. In the AQA mark scheme you need to think about these since they form part of Area 4: Conclusions, evaluation and presentation.

In most investigations where you are dealing with quantitative data you have to think about the twin issues of reliability (can the results be replicated), and validity (does the survey measure what it was intended to do).

If you have designed a good investigation, then hopefully you will have anticipated various types of errors that can affect reliability or validity. Your anticipation, perhaps through developing a pilot survey, will have reduced the errors and/or their impacts. Nonetheless, *errors will be present* and they need to be analysed and assessed. All research, even at the highest academic level, is flawed and has errors of some description. We will return to reliability and validity in Section 5.10 when we think about the evaluation.

There are lots of different ways to think about errors. The next two pages will introduce you to some categories and classifications.

It's not all about limitations

Be careful not to confuse limitations with excuses as to why something hasn't worked or turned out as you expected. At this level, limitations can take you down a blind alley since it often looks like you have not really given enough consideration to either design or methods. It is also wrong to think that identifying problems with your design and methods somehow represents failure.

There are some valid limitations that can be identified. These include issues with safety or site access or having to use an alternative piece of equipment as the resources did not exist to purchase a more expensive version. You should comment on how these made a difference to your investigation and how these limitations affected the *strength* of conclusions you were able to make.

Design considerations and the introduction of sampling errors

This is the simplest type of error to be aware of, although it is often difficult to control because of the time and effort required to collect bigger samples. On page 31 you looked at sample size, which is a factor that has implications as to whether there are differences between the observed sample and the rest of the population. A small or biased sample is likely to produce a result which can lead to an interpretation that is biased or skewed, not reflecting the population as a whole.

If the sampling procedure (or sample frame) is defective, then it is likely to introduce sample bias. Look at Table 16, which gives descriptions and considerations of different types of sampling. You may also encounter other types of sampling in your literature review, such as opportunistic sampling that need attention. They are more than likely flawed because their very nature is opportunistic or pragmatic to gain information and therefore likely to be unrepresentative of the population.

Table 16 Considering different sampling methods.

Method/procedure	Design description	Considerations
Systematic sampling	Samples are chosen in a systematic, or regular way, e.g. every ten minutes, or every hour, or every seventh person. This is used when the environment or population has an expected environmental gradient or change (spatially or temporally), but the degree of change may be uncertain.	Can give good coverage (spatially of an area) and is straightforward to design but has the potential to miss areas when surveying along particular points or lines (transects), which will lead to gaps and an under- or over-representation of certain groups or features in an area.
Stratified sampling	Samples are taken at pre-determined places or times based on an understanding of the study area in terms of the groups, individuals and sub-groups. This is used when the environment or population has an observed environmental gradient or change (spatially or temporally), and the expected change can be used to inform the sampling procedure.	This approach reduces the potential for bias in areas of variation, but the sampling design frame needs to take account of the underlying characteristics of the area or population in order make the correct selections. In some instances, it can be impossible to get data on groups in order to stratify the sample (e.g. ethnicity of tourists to a town).
Random sampling	This is sampling using random numbers to generate times and/or co-ordinates for when a sample should be taken. This is used when the environment or population has no known environmental gradient or is thought to occur at random. It can also be used when there is no assumed knowledge of the population.	This sampling approach should minimise any elements of human bias and therefore sample error. However, random sampling can leave gaps in the sampling design frame, or lead to an undesirable clustering of points. It can also be time consuming to undertake compared to stratified or systematic sampling.

Measurement errors

These are mistakes made when collecting the data, such as problems reading a clinometer or thermometer. A good investigation will usually try to minimise these through practice and careful procedures. Repeat measurements will reduce errors and anomalies.

Operator errors

These relate to the different interpretations as data is collected by different people. They are especially relevant with any bi-polar scoring systems, or where judgements are being made using a numerical scale. If you are working in a group for instance, pre-calibration may help, or averaging results before interpretation.

Questionnaires and interviews: response and non-response errors

If you are carrying out questionnaires and interviews then there is a special type of category that may be worth consideration. The process whereby ideas are exchanged and recorded during an interview or questionnaire is subject to error. For example, questions can be misunderstood, or respondents may feel pressured in responding to the researcher's own ideas.

Non-response errors are linked with interview and questionnaire sampling – biases associated with who did or who did not respond.

Evaluating data collection

Look at the following pitfalls which are often written about as part of an evaluation of data collection.

- Limited and vague reflection (such as 'sample size wasn't big enough' or 'there were problems with the clinometer').

- Focus on limitations in methods rather than design (such as 'questionnaires could have been better sequenced' or 'pedestrian count was difficult to do').

- Suggesting that the study is without flaws and errors in design (such as 'the results show this, … I'm sure that this is the correct answer').

Write some comments to explain how these are not really relevant or should be modified to suit a GCE investigation.

My progress

First, categorise your appraisal of both your design and methods of data collection in terms of errors using the table below.

Now, reflect which had the most impact on your results. Circle or highlight those which you think at this stage are most influential in affecting your conclusions and write notes on them in the box below.

	Design	Method
Limitations		
Sampling errors		
Measurement errors		
Operator errors		
Response and non-response errors*		

* If relevant

5.6 Data presentation

Data presentation (or representation) allows the reader to see what information you have collected. It helps begin the geographical narrative that you are telling as part of your investigation.

Raw numbers and tables of data (especially large and complex ones) on their own are difficult to understand and interpret; they can also be daunting. Graphs can tell a story visually rather than by using words or numbers, and can help the reader understand the meaning in the data rather than the technical details behind the numbers. Graphs can highlight patterns and trends in the data. This again helps in the understanding of the data, but also helps the reader to spot any anomalies or irregularities. They can, at a glance, help to determine the reliability or trustworthiness of the information presented.

Remember cartographic refers to maps, whereas graphical refers to graphs. You can use a combination of both of these on the same diagram.

The following two pages will provide you with practical ideas on how to make your data representation achieve the top marks.

Good graphicacy

Graphs can easily be produced using readily available spreadsheet software but you need to be very careful in selecting the most appropriate type of graph to display data. It's not always the first or the 'flashiest' in the range of types available. In general, it is safer to avoid 3D graphs as they can make the data appear too cluttered and hide some variables or data behind each other. There are certain elements that should appear on *all* graphs:

- titles for axes
- labels
- graph title
- plotted data
- reference to the data source.

There are several other top tips which demonstrate good graphical practice.

- The chart area defines the boundary of all the elements related to the graph including the plot itself and any headings and explanatory text. It emphasises that these elements need to be considered together and that they are separate from the surrounding text.
- The x-axis is the horizontal line that defines the base of the plot area. Depending on the type of graph, the x-axis represents either different categories (such as years) or different positions along a numerical scale (such as temperature or income).
- The y-axis is the vertical line that usually defines the left side of the plot area, but if more than one variable is being plotted on the graph then the vertical lines on both the left and right sides of the plot area may be used as y-axes. The y-axis always has a numerical scale and is used to show values such as counts, frequencies or percentages.
- If the graph you are presenting is based on data from another source then you should acknowledge this somewhere within the chart area or title.
- The choice of shading and/or colours is also important – see the activity on page 52.

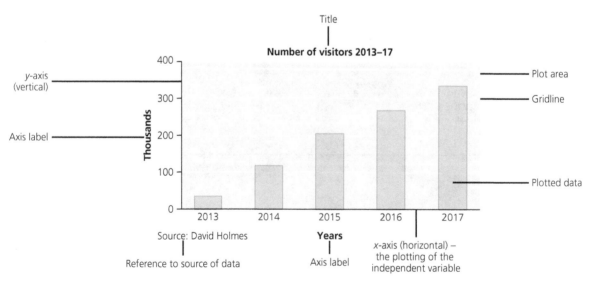

Figure 24 An example of good graphicacy.

Benefits of GIS – geo-spatial location

GIS has lots of benefits, not least that it can show the spatial relationship between different pieces of interrelated data. This really helps in the deconstruction of data and links or connections that might be present.

The example below shows pedestrian fluxes (flows) along a transect. Take note of the attention to detail in terms of:

- The selection of the base map so that street names are visible to locate sampling points.

- The choice of blue as the colour for proportional symbols as there is very little overlap with that colour on the base map. That means they stand out.
- The size of the blue proportional symbols has been scaled so that differences between locations are clear. They have also been put into categories, rather than displayed as actual numbers which makes comparisons more obvious.

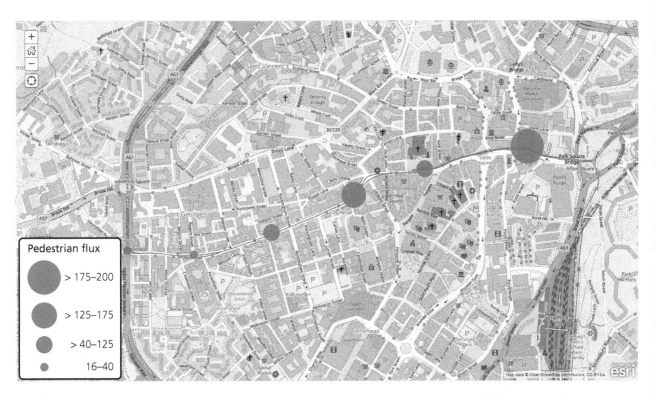

Latitude	Longitude	Pedestrian flux
53.38325	–1.46373	200
53.38258	–1.46899	125
53.38181	–1.47262	150
53.38071	–1.47677	90
53.38006	–1.48072	40
53.38020	–1.48410	16

Figure 25 A GIS map for Sheffield showing pedestrian flows. Note the table of data which is used to plot the GIS information, according to latitude and longitude. © OpenStreetMap contributors.

Look at this presentation of histograms which shows the effect of different shading and colours. The choice of these is important because some will make it easier for the reader to interpret.

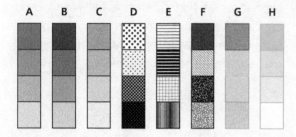

1   Which of columns A–H is clearest and why?

_____

2   Which columns have problems associated with either their choice of colours or shading?

_____

3   Look at columns A–C. Does the grey shading order matter? Which of these is the clearest?

_____

My progress

It is time to try to audit your possible techniques and approaches. There is no minimum number here, but there is a danger that sometimes too many are used.

● Indicate the most influential in relation to your aims by using an (*).

● Indicate whether they are going be presenting primary data (P) or secondary (S).

	Justification
Cartographical	
Graphical	
Other, e.g. tables, photographs	

Is there any data that you are NOT going to present? If not, why not?

Remember that analysis is about making sense of the data and information you have collected so that someone reading your work can understand your reasoning and judgements. It is also about telling stories with data and information.

Critical analysis is a heavy-weight of the AQA mark scheme. It is also an area where there is often an opportunity for improvement. When you have quantitative data it is recommended that you follow a sequence that forms part of the analysis process. This will likely include:

- an initial description of any main patterns and trends
- further analysis and confirmation of the patterns and trends using data or information from the relevant tables, charts or cartographic evidence
- identification of anomalies or exceptions which deviate from the main patterns and trends.

It is probably useful to think of presentation providing a natural lead into analysis. There is often an overlap between presentation and analysis, for example a box-plot graph shows the spread of data statistically (see Figure 26).

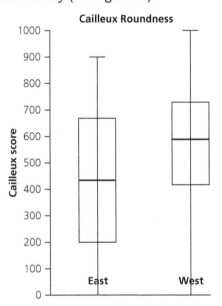

Figure 26 A box-plot graph.

You will probably integrate your presentation and analysis, rather than having them as separate sections. This makes it easier for the reader to refer to the relevant charts and diagrams.

A range of quantitative analysis techniques available

With a quantitative (numerical) set of results, the following checklist may be useful.

- What is the range of values within the data set?
- Where are most of the values concentrated (i.e. is there any clustering)?
- Are there any clear gaps between the concentrations?
- What shape is the distribution of the data values?
- Are there any extreme values (which may include anomalies and/or outliers)? How far separated are these from the normal range of data?

Table 17 below provides some different statistical methods. Remember that spreadsheets can really help deal with large secondary data sets. Various functions like FILTER and SORT can quickly rank and order data, making analysis more meaningful.

Table 17 Different statistical methods.

Statistical function	Identification of method
Test for differences in sets of data	Chi-squared test
	Student's t-test
	Mann-Whitney test
The linkage (correlation) between two sets of data	Scatter graph and line of best fit
	Calculated line of best fit (e.g. using Excel – 'Trendline')
	Spearman's rank correlation
Summarising and describing data – central tendency	Mean
	Mode
	Median
Extremes and thresholds	Use functions in Excel: = MAX, = COUNTIF
Spread, dispersion and range of data	Standard deviation
	Quartile deviation and interquartile
	Range

The 'conditional formatting' command available in a spreadsheet will also help you to show patterns and trends in your data more easily.

In Table 18, where green and yellow colours have been used to show satisfaction scores.

Table 18 Displaying satisfaction scores.

Scores related to visitors' impression of a city rebranding scheme	4.25	4.75	5.00	4.75	4.75
	Overall 'imagability'	Ease of access	Quality of design	Ease of movement	Range of facilities and shops

Always remember that your analysis of data should focus on the geography, rather than just 'doing' the maths.

Using other statistics

Descriptive measures of difference and association (such as Chi-squared tests or t-tests), plus relational statistics (such as Spearman's rank correlation or lines of best fit) are also likely to be something you will want to consider. However, there are some precautions to think about:

1 Check that you have enough data to perform the test(s) so that it is valid.

2 Choose the right test for the data that you have. Don't assume that one test will be better than another or get more marks. It is a question of fitness for purpose.

3 Don't perform a statistical test just for the sake of it, when it is, for instance, obvious that there are differences between sites or locations. You should instead think about using a test when the other evidence is inconclusive.

4 Interpret the test carefully, referring to confidence levels. Remember the difference between correlation and causation.

There is a lot of information online and in books that can provide you with a step-by-step guide for undertaking this analysis. And remember that there are several online calculators that could also be useful.

Qualitative analysis

Analysing qualitative data remains an area of weakness in many fieldwork investigations, yet there is much you can do in this regard, and you shouldn't see this data as in some way less technical than doing the procedural maths associated with quantitative data. Annotating photographs is an obvious example of analysis but there are several other techniques:

- Linkages – mapping links between findings, for example using a concept map.
- Coding – extracting meaning from lengthy open responses.
- Polarising – analysis of positive and negative impacts from text.

- Polar scaling – semi-quantitative analysis of negatives and positives, based on context.
- Theming – identifying bigger themes and linkages in responses and text.

> ### Extracts from an interview
>
> 'We have a very strong relationship with the staff and feel very supported before and during the visit.'
>
> 'There is measurable excitement in the children as we approach the gates.'
>
> 'X represents excellent value for money and is very accessible, which are key factors in choosing to come here.'
>
> 'There is always ample car parking and the toilet facilities are generally clean.'
>
> ### Summary
>
> This visitor has been coming to X for around ten years. They have used other places and providers but really value the activities that X is able to offer. The interview revealed that the staff at X make the 'fun' and 'adventure' aspects of visits very real and are able to show how the park has invested considerably in visitor amenities and facilities.

Figure 27 Example of a summary from an interview at a local tourist attraction.

Making more of images – the place narrative

Images offer lots of potential for analysis as part of your fieldwork investigation. Rather than annotation, you might consider a more discursive approach, creating a narrative. In many ways this is using a written approach to link together ideas and processes. Perhaps working out what the area used to be like and how it might change in the future.

See Figure 28 on an area of Leeds.

This photograph is of a zone of urban land use with a canal waterway dissecting the image. The scene appears to show historic buildings, likely Victorian with an industrial purpose. In much of the picture there is evidence of change: improvement and modernisation with a resulting pattern of gentrification most likely. This will have increased the desirability of the area – becoming an attractive waterfront area with flats and/or offices in the foreground near the canal barge. The process of re-urbanisation may also be happening. Such improvements tend to reduce the socio-economic mix, and create places and spaces where some residents will be excluded because of housing affordability.

The place is continuing to change and undergo improvements, but these are driven by private sector development which is far from resilient. A downturn in the economy, house prices and place vitality will create an uncertain future in terms of further enhancements to housing stock.

Figure 28

A framework for qualitative analysis

Table 19 gives an example of a theming framework using three different sources of information – in the table these are found along the top row. Theming is a way of bringing together a number of approaches which can help identify connections between concepts, data and information collected from both primary and secondary sources. Theming has the advantage that it goes beyond just counting of words and phrases; instead it attempts to synthesise patterns within the evidence itself.

Table 19 An example of a theming framework.

	Concept (1): Changing nature of shops and services	Concept (2): Gentrification leading to high cost of house prices	Concept (3): Seasonal and low-paid employment
Respondent – interview (1)	Has seen a decrease in the number of affordable cafés and other places to eat [2:33]	Knows many people who cannot afford to move to a bigger house, so they are worried about having a family [4:27]	
Local plan – document (2)		Wants to encourage more low-cost housing and identifies possible sites [page 23, para 3]	The authority is working with local businesses to pilot a living wage scheme and better pension provision [page 75, para 2]
Forum – internet research (3)	People are worried about more outdoor shops and lack of high street balance [Dave9066, 3rd comment]		[Janet K, 6th comment] is concerned about lack of flexibility and low wages for part-time mothers.

Effective critical analysis?

Now that we have considered some key ingredients of good analysis, read the analysis extract below and answer the questions that follow on a separate piece of paper to decide whether it is effective or not and why.

> Graph X shows the correlation between the average grade of stone size and the distance from the water's edge. Overall, all of these recordings have a similar trend and pattern, in that at all three sites the stone roundness increases as the distance from the water's edge increases. Sites 1 and 3 are very similar as they both have the same increases and decreases at the same four points. For instance, site 1 increases from 3 to 4 when looking at 10 m to 30 m, however it then falls to 3.7 at 40 m and then increases to 4.3 at 50 m. Site 3 is very similar as it also increases from 3.4 to 4.3 when looking at 10 m to 20 m.
>
> When looking at the theory of stone roundness on a beach I would expect to see the stones increase in roundness as you increase the distance from the water's edge. This is because there are less erosional processes taking place at the back of the beach which the water reaches less frequently. However, on a beach that has beach nourishment you wouldn't expect to see a trend, as the sediment would be all moved around and not in a specific pattern of shape or sizing.

1 What do you think of the overall use of language, style and structure?

2 Should the first paragraph have had more description of data or less?

3 Is the geography correct in the second paragraph? If not, what is wrong?

4 How and where could they have explored links to wider theory or the research literature?

5.8 Making links and connections, and keeping to the word limit

At this stage some of your analysis and interpretation is drawing to a close but there is an important task still to do: find links and connections between the areas of geography that you have been considering. Often such links can be demonstrated with personalised diagrams and mind maps, rather than more text. Diagrams may have an additional advantage in that you can show the strength of linkages as well as keeping down the total number of words.

Another important aspect to consider here is the broader geographical context. In other words, how your work is of interest and shows links beyond the locality that you have studied.

Using diagrams to make connections

Many geographers tend to be visual learners, therefore, the use of diagrams can help us communicate ideas about what we have found in a more concise and meaningful way than expansive text. There are several methods by which data and information from your investigation can be synthesised and reported:
- presenting information in tables and charts and then colour-coding or ranking
- developing your own measures and scales to summarise data
- using annotations to interpret the main features of data, highlighting trends and anomalies.

However, perhaps the most useful is using personalised diagrams to show geographic linkages between pieces of information. You have already done something similar when you began to plan your write-up in Section 5.1, page 38. Figure 29 is a different example which summarises the information from a series of coded interviews investigating the identity of Milton Keynes. This attempts to make links beyond just the interviews, while considering other parts of geography that might be relevant.

Unlocking the broader geographical context

There are many opportunities to link to the wider geographical context and to use your experience to extend geographical understanding. These opportunities include ideas around risk, resilience and identity. There are 13 of them in total. Table 20 is one example of how you could use these concepts as links to the broader geographical context. The link is highlighted in green.

Table 20

Concept	Linkage
Resilience	A study of local catchment hydrology and flood response after a storm. Future resilience with climate change and changing storm intensity and rainfall patterns.

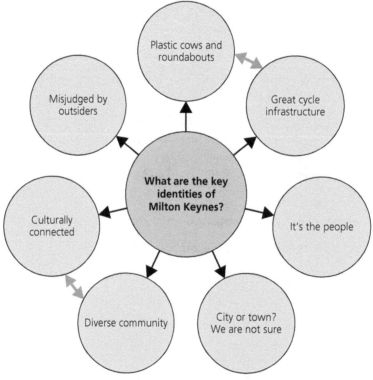

Figure 29 The identity of Milton Keynes. Note that possible links are shown with the blue arrows.

Managing the word limit

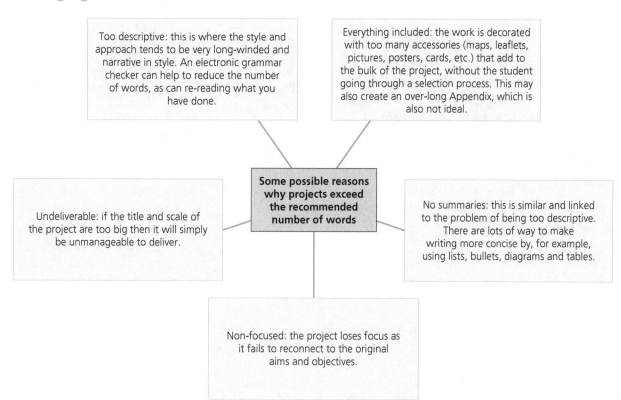

Figure 30

Remember that the NEA is recommended to be between 3000 and 4000 words in length. Often candidates exceeded this considerably. All awarding bodies have stated that students should produce investigations that are concise and focused throughout. More words do not necessarily mean more marks. Remember there is not a specific penalty for over-long work, however the person marking your NEA may decide to lower their score based on the AQA Area 4: Conclusions, evaluation and presentation where work needs to be 'balanced and concise'. Excessive length is self-penalising in terms of time taken and has a tendency to lack focus or clarity.

If you are way over the recommended number of words, then it is more than likely you are doing something wrong. There are several possible causes outlined in Figure 30.

Correlation versus causation

When we see a link between an event or action, what sometimes comes to mind is that the event or action has caused the other. This is not always so: linking one process, action or activity with another does not always prove that the result has been caused by the other.

This is at the centre of the causation versus correlation discussion and it may be very relevant for your fieldwork investigation. Read the following definitions:

Causation

An action or occurrence that can cause another. The result of an action is always predictable, providing a clear relationship between the actions and the result, which can be established with certainty.

In general, it is extremely difficult to establish causality between two correlated events or observations.

Correlation

An action or occurrence that can be statistically linked to another. The action does not always result in another action or occurrence, but you can see that there is a statistical relationship between them.

Correlation can be easily calculated and demonstrated through statistical tools, such as Spearman's Rank or other correlation coefficients.

Causation or just correlation?

Read the following five examples. They all *may* show statistical correlation, but is there any causation linkage? In other words, has one variable had an effect on the other? Can you give a logical, geographical explanation for any of these?

1 Sales of ice cream and the amount of crime in a city.

2 Number of priests in America and alcoholism.

3 Number of mobile phone masts in an area and the birth rate.

4 Density of homeless people and incidence of crime.

5 Shoe size and IQ scores for a group of students whose ages range from 10 to 15 years.

1

2

3

4

5

My progress

Let's revisit the specialised concepts that we looked at briefly in Section 5.3: Writing the introduction and preliminary research. You will remember that there are 13 in total, namely: causality, systems, feedback, inequality, identity, globalisation, interdependence, mitigation and adaption, sustainability, risk, resilience and thresholds.

Use one of the synoptic concepts and write a short paragraph on how your project can be linked to it.

5.9 Writing conclusions

A conclusion is, in some ways, like an introduction. You should restate your aims or questions and summarise your main points of evidence for the reader. Critically, conclusions must draw on evidence presented in the rest of the fieldwork investigation. You may find that the question is only partly answered as the data or information collected only provides contradictory evidence or it only applies to part of the area you were investigating. You could say – 'in some parts of the region, there is acute deprivation ...' and, later, 'while in other ways deprivation is more hidden and less obvious, for example, in area X'.

Conclusions can be tricky since they require drawing across the information in the NEA, rather than summarising single pieces of data or ideas. There can be several problems with conclusions:

- They don't effectively summarise information into distinct statements and ideas. This is where an abstract (see pages 65–66) or a brief summary, may help.
- They are too long and/or too vague.
- They don't draw on the evidence found in the analysis in the rest of the work.
- They are limited to problems, rather than how shortcomings might have impacted on the strength of individual conclusions.

Remember that conclusions are built around a series of individual stages or components, rather like a jigsaw puzzle.

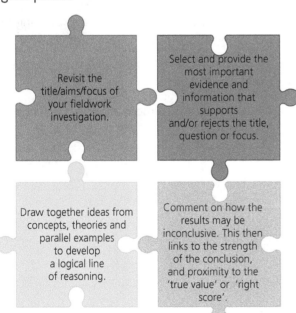

Figure 31

Trying to get to knowledge and wisdom

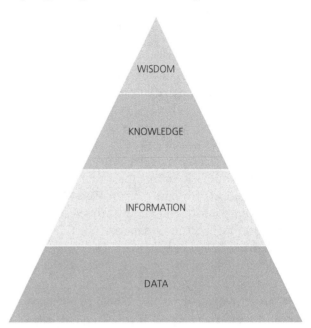

Figure 32 An information hierarchy.

Look at Figure 32, which is often called the information hierarchy. This could be applied to your fieldwork investigation:

Table 21

Wisdom	Showing a deep understanding about the geography: an appreciation of the what, why and how and the what if, in light of future change
Knowledge	Recognising the significance of individual components, conceptual linkages and frameworks. 'Know-why' in relation to processes.
Information	Analysis of data and information to create order and structure.
Data	Primary and secondary fieldwork data – raw and unprocessed. Signals nothing on its own.

Moving towards the top of this pyramid should pay dividends since the mark scheme is rewarding skills of wisdom and knowledge. A conclusion then can offer the 'know-why', and 'why-is' type of information that shows evaluated understanding.

Two models for a conclusion

There are lots of different ways to approach structuring a conclusion. One strategy is to think about ranking conclusions, based on how strong the evidence supporting each one is in turn, for example:

Nevertheless, this investigation does suggest a number of possible conclusions. They are presented in order, with the most secure, going to the least secure.
- Perception and awareness of crime does not match the calculated risk.
- That people in Truro have a number of misconceptions regarding the actual geographical distribution of the risks of crimes.
- Secondary information about crime can be misleading in terms of patterns.
- There is no clear correlation between deprivation and risk of crime at a spatial level.

An alternative approach is to use bold to highlight the most important finding or your summary idea.

The perception and awareness of crime does not match the calculated risk in Truro and there are misconceptions about the actual geographical distribution of the risks of crimes. This work shows that secondary information about crime can be misleading in terms of patterns and there is no clear correlation between deprivation and risk of crime at a spatial level.

Messy and complex geography

Often you might find your conclusions are a bit messy, in other words the results and evidence are inconclusive. It's fine if you say how far you think a conclusion might hold up and don't be afraid to use words such as 'partial', 'tentative' or 'incomplete'. Your conclusions may also disagree with a model or concept that you introduced at the start of the investigation. Don't worry about any differences. It may be you who is 'right' for this place, at this time, based on the evidence that you collected. Be confident in your writing and approach. It might be good practice to draw out the differences between the model or concept and the actual data collected and possibly consider why these differences might have occurred.

When starting my investigation, my research led me to the view that the current government policy on affordable housing in London was not fit for purpose. Many of the affordable housing schemes had simply failed.

However, the primary evidence that has been collected, together with the in-depth analysis of secondary data, has thrown up an alternative and opposing perspective. That is, that the current government policy on affordable housing is, at least in part, both working, and fit for purpose. In particular, the primary evidence based on questionnaires, makes the case that many people are willing to accept that the government are providing safe homes. The majority of respondents also think that the government has adopted a policy of trying to give preferential treatment for local residents where there is a shortage of housing …

You might also consider about introducing possible new or expanded ways of thinking about the focus for your NEA. You can offer a new insight and creative approaches for framing or contextualising the research problem based on the results of your study.

My progress

Prepare your conclusion in summary notes below. Use numbers to rank your most important findings.

5.10 Writing the evaluation: reliability and validity

In Section 5.5, page 47, you considered the ideas of reliability and validity as part of an evaluation of your methodology and design. For the AQA mark scheme it is good practice to reflect on the accuracy and reliability of the conclusions. This forms the main body of the evaluation component (Area 4), where you are expected to appraise the quality of your own evidence and the strength of your conclusions. All in all, this part of the write-up is quite tricky.

'Reliability' and 'validity'

When you think about your critical evaluation, two big questions should always be in the back of your mind:

1 How **reliable** are the descriptions, findings or data?

2 How **valid** are the assertions, assumptions, arguments or conclusions?

> ### Reliability
> The consistency or reproducibility of a measurement. A measurement is said to have a high reliability if it produces consistent results under consistent conditions.

Central to the quality of the critical evaluation is an understanding of these measurements in the context of validity.

> ### Validity
> How well a scientific test or piece of research actually measures what it initially sets out to do.

This section therefore needs to be written considering both these terms. You might want to link your comments about reliability and validity to conclusions and express them in terms of confidence, as shown below, related to what you set out to find.

Very confident	Confident
Somewhat confident	Not confident

A checklist for reliability

In order to measure the reliability of your investigation and its evidence (both primary and secondary), several questions can be asked:

1 Is the aim of the investigation clear and appropriate? Does your study describe how the data was collected (sample and design) and is there any previous research into the reliability of this data collection approach?

2 Was the study sample appropriate? This includes a consideration of both sample size and sample design as well as methods used.

3 Does your design minimise **random error** in the methodology? These include random errors and **systematic errors**.

4 Are the results meaningful? Does the statistical test generate an outcome at a 95 per cent confidence level?

You can use and adapt these questions to give structure to your critical evaluation. One advantage of group data collection might be that you can think about interrater reliability. In the context of your investigation that might be simply comparing your data at a similar site to someone else's, and confirming agreement. High agreement means that you should be more confident in your outcomes. You might also find that other empirical research and data also support your findings.

Thinking like a geographer

A valid conclusion is only supported by reliable data using a valid method and based on sound reasoning. Some key ideas here are:

1 *Be objective*: have you considered the different sources of error (Section 5.5) and the impact on outcomes?

2 *Be critical*: how have you judged that your conclusions are valid?

3 *Be ethical*: how successfully have you embedded ethical thinking (see page 63) and did those impact on your outcomes?

Adapted from: Source FSC 2016 OP171

My progress

The language of your critical evaluation is going to be important. Terms such as 'a lot', 'pretty good', 'close' or 'very short', do not have a place in high quality geographical evaluations since they lack real meaning. These examples are relative terms – words whose meaning can change depending on what they are compared to or the context in which they are used.

Review the list of phrases below that could be useful as part of your critical evaluation. Indicate those which are directly relevant with a tick, and then think about how they could be adapted to be included as part of your evaluative write-up.

Example critical statement	Useful?
The evidence points to a conclusion that is secure in terms of …	
Inevitably, several crucial questions are left unanswered by this investigation. They include …	
Despite my inability to … I was greatly interested in …	
As a sixth form student I feel that this article/book very clearly illustrates …	
The tables/figures do little to help/greatly help the understanding …	
This explanation has a few weaknesses that other researchers/authors have pointed out …	
As it stands, the central focus of X is well/poorly supported by its empirical findings …	
Less convincing is the broad-sweeping assumption/generalisation that …	
Nonetheless, other research data tend to counter/contradict this possible trend/assumption …	
When considering all the data presented … it is not clear that the low scores of X, indeed, reflect …	
In addition, this research focus proves to be timely/especially significant to … as recent government policy/proposals has/have …	
Over the last five/ten years this topic X … has increasingly been viewed as 'complicated' …	
This argument is not entirely convincing as …, furthermore it justifies the …	

5.11 Consideration of the ethical dimensions of field research

It is a requirement of your fieldwork investigation to consider the ethical aspects of how carrying out your project may impact on people and/or the environment. Your job is to identify and manage such impacts. This should be an integral element of high quality data collection and design.

The most common ethical dilemmas in human geography are around participation, consent and the safeguarding and confidentiality of information.

In physical geography, the main ethical considerations are around consent and access to study sites and potential damage. In most instances, careful planning should effectively reduce the impacts and effects to a minimal level. This could be done as part of a risk assessment.

Students have a responsibility not only towards others and the environment but to take care of themselves both physically and mentally when carrying out their research.

Why are the ethical aspects important?

The AQA mark scheme includes ethical considerations as part of Area 4: *'show an understanding of the ethical dimension of field research'*. There are other important aspects of ethical consideration.

- It protects the rights of the individuals and communities that are affected by the research, as well as the environment where the research is taking place in.
- It meets the growing public demand for accountability, for example, the 2018 GDPR (General Data Protection Regulation).
- It shows that you are serious about your investigation and want to carry it out properly, professionally and correctly.

Some principles of ethical behaviour

Several researchers recognise the following principles of ethical behaviour.

1 *Do no harm*: Safeguard against anything that could harm participants or the environment in your study.

2 *Privacy and anonymity*: No identifying information about an individual should be revealed. Seek permission from participants.

3 *Confidentiality*: Information collected should not be given to anyone else.

4 *Informed consent*: Participants are informed of the nature and extent of the study.

5 *Truthfulness and accuracy in reporting data*: Data will not be created or falsified.

6 *Intrusiveness*: Remain a neutral researcher.

7 *Data interpretation*: Researchers should use their data to fairly represent what they see and hear. Don't misinterpret the data collected to present a picture that is not supported by the evidence.

Ethical considerations

Usually ethical considerations will form part of a methodology but in conclusions and/or evaluation you should also consider how it might affect the results. For example, assumptions about other cultures or about a particular cultural group may bias our perception. They can impact on objectivity including issues of stigma, stereotyping, and even discrimination.

The amount that you write will be dependent on the nature of the research. Generally, a paragraph will be sufficient to explain your actions and reasoning to the reader, although it may be longer for a human- rather than a physical-based piece of work.

Example of ethical practice: interview

For face-to-face interviews, participants (respondents) can be informed of your ethical planning at the start. Table 22 is an extract from notes for that could be used at the start of the interview.

Table 22 An introduction to a face-to-face interview.

Objectives/ timing	Ethical brief
Introduction (3 minutes)	Nature of research and how it will be used: <ul><li>This research study forms part of an independent review of X.</li><li>Recording will take place for recollection purposes and to obtain quotations.</li><li>Anything said will be treated as confidential and anonymous. Your personal data will not be passed on to anyone else.</li><li>Think of this as an informal chat. I'm interested in your own reflections on X and how you feel about the planned project at Y.</li><li>The interview should take around half an hour.</li><li>I am happy to send you a copy of the final research when it is completed if you want me to.</li></ul>

Example of ethical practice: physical geography

Table 23 is an example of ethical considerations from a physical geography investigation.

Table 23 The ethical considerations of a physical geography investigation. Source: Field Studies Council (2018).

Method	Ethical and socio-political considerations
Clast analysis (axis a length and shape of sediment – Powers Index of Roundness) on storm beach	Consider access to the site, and gain permission by any relevant land owners.
	No beach material is to be taken away from the site.
	Ensure that any trampling or equipment, e.g. tape measures, ranging poles, does not disturb any of the vegetation that may be present on the storm beach.

Evaluating the ethical issues

Below are some extracts from an NEA where a student is writing about ethical considerations for a proposed questionnaire. Look at the highlighted sections. Those in green are examples of good practice, those in blue are more questionable. Take a few minutes to review and work out how it could be improved.

When designing my questionnaire I was aware of asking questions that could be seen as too invasive, so I asked my friend to take a look first.

My pilot study revealed that people thought I might work for the council or even the police so I made sure that I introduced this as being a personal and private piece of study research.

I showed people my college ID card to demonstrate my age.

I gave people my personal mobile number if they had any follow-up comments that they wanted to talk to me about.

I asked people if I could take a photo of them and record comments on my mobile phone.

I told participants that I would post images of them on Twitter as part of my research.

I removed the stones from the beach and took them home so that measurements would be easier.

I was very careful not to damage the sand dune plants and to reduce my trampling impacts.

My progress

Write a short paragraph which covers your ethical considerations. You can then use this directly in your fieldwork investigation.

5.12 Reviewing example coursework extracts and thinking about an abstract

Here are three extracts of writing for different parts of fieldwork investigations. There is a commentary on each one, together with a proposed title or focus to give the work a context.

1 Extract from the Introduction and preliminary research: Flows of people and investment into an area of NW London

Shifting flows is the immigration into an area that leads to the change in its demographic and cultural characteristics, sometimes in a short period of time. Investment causes economic change as well as influencing social inequality. Over the years, Wembley has overseen a large amount of change from shifting flows of people and investment, causing it to develop to some extent.

This comes from the start of an introduction (purpose). While it's clear what topic the student is intending to study, the language is somewhat vague and lacks technical vocabulary. The timeframe has not been established, other than 'over the years'. It might be good to see at this stage some explicit reference to some literature research, combined with a model, concept or even comparable study. There is also very little to ignite interest. Why is this an interesting topic? Why, in particular, has this student chosen this topic? Why might it be relevant in NW London?

Although it is not compulsory to do an abstract, it may help to think about doing a short and succinct summary of your whole investigation. The abstract is usually no more than one paragraph of five to seven carefully developed sentences that provide the reader with what the paper is about, what the document intends to do, what methodology was used, what it will conclude and what action (if any) is desired from the reader.

This is generally a very competent piece of writing, although it falls short in terms of its comments about the conclusion and the evidence collected; for example, it would have been stronger if it had stated numerical data to support the result.

2 Extract from the methods of critical analysis: Impact of people on the coastal sand dune ecosystem (psammosere)

This secondary data locates the blowouts in the Braunton Burrows psammosere. Two patterns are clear, there are no blowouts close to the honey pot site, and the further away the blowout is from the honey pot site, the smaller it is. The first trend can be explained as the exclusion zone restricts human interference in the psammosere and hence reduces erosion. The second trend can be explained, through the attraction of the local beach café and amenities. Tourists aren't willing to walk too far away from the honey pot site, and so blowouts further away from the HPS are visited less frequently and so they experience less erosion. Strangely, a large blowout that I encountered along my transect from 100 m to 150 m hasn't been recorded in this secondary data, however if I was to use this blowout, it would still conform to the trend as I recorded the blowout as having a circumference in excess of 260 m.

This is a successful example of framing data and results for a target audience or another reader. The writing is clear and technical where it needs to be. There is for most of the extract use of data as evidence to support the line of reasoning, although there perhaps should be more at the start. The reader is presented with ideas and clear explanations; exceptions are also considered, producing a clear and logical text. The student is deconstructing the data to show connections.

3 An abstract: A study of crime in Cambridge

This study is an investigation of the geography of crime in the city of Cambridge – the incidence of assault, burglary and theft of motor vehicles. Crime is a major aspect affecting quality of life and blights communities. There are two main objectives: 1) to establish the spatial distribution of perception of crime, and 2) to establish the pattern of the actual distribution of crimes. The investigation is based on questionnaire data, local police records and secondary census and insurance information. The results show, in particular, the gap between perceived and actual risks of crime in different parts of the city.

My progress

Here is a checklist to help you think about the writing of your own personalised abstract. Abstracts are usually written in the present tense, but you might refer to prior research in the past tense.

- Motivation: why have you undertaken the research – *why do we care* about the problem, process or issue?

- Problem statement: what problem, process or issue *are you trying to solve*?

- Approach: what *methodology did you use* to collect the data (primary and secondary)?

- Results: *what's the answer*, quote using evidence and be precise.

- Conclusions: what are the implications of what you have found? Is it important, different or does it agree with other research and is what you expected?

Remember the choice is yours as to whether you use an abstract in your completed fieldwork investigation but it certainly shows some aspects of good practice. It is also something that is often required in higher education, for instance at university. Abstracts form the opening page of a research report, before an index or table of contents.

Complete your abstract using a maximum of seven sentences and build them into the answer box, using a scaffolding approach.

Motivation: why have you undertaken the research – why do we care about the problem, process or issue?

_____

_____

Problem statement: what problem, process or issue are you trying to solve?

_____

_____

Approach: what methodology did you use to collect the data (primary and secondary)?

_____

_____

Results: what's the answer? Quote using evidence and be precise.

_____

_____

Conclusions: what are the implications of what you have found? Is it important, different or does it agree with other research and is what you expected?

_____

_____

1

2

3

4

5

6

7

References and bibliography

Almost all academic writing will need a reference list. In carrying out your research for the NEA you will have seen in many of the books and articles that you have used that they contain footnotes (at the bottom of the page) or endnotes (at the end of each chapter or section) in which authors acknowledge other works they have used as part of their research. This makes clear that the author is not claiming this work as their own research.

Similarly, at the end of the book there will be a bibliography. This is somewhat different from footnotes or endnotes. A bibliography is a list of all the books that authors have referred to during their research, even if they have not quoted from them or taken ideas and information from them.

References

A comprehensive list of the sources that you have referred to in your writing so that the reader can follow up any source you have referred to.

Bibliography

A list of sources at the end of your writing, including sources you have not referenced, and sources you think readers may want to follow up on.

In the context of a geography NEA, referencing is sufficient and good practice, but a bibliography may provide more evidence of additional reading and research.

The purpose of referencing

Referencing in the NEA is important for a number of reasons:

- You need to show where you obtained the knowledge, concepts and ideas that you are using.
- You cannot claim that ideas and knowledge are your own when you have taken them from other people's work – that is plagiarism (see page 69).
- It is an academic convention, and you are producing an academic piece of work.
- It is good practice for the future, as higher education institutions will expect it.

Referencing systems: Harvard versus footnote

There are two commonly used referencing systems that would be suitable for a geography investigation.

1 Author/date or Harvard system

This references an item of literature using the author's name and year of publication. These are then listed at the end of a report. (See Table 24 for examples.)

Table 24 Referencing styles.

A book	Phillips, R. and Johns, J. (2013) *Fieldwork for Human Geography*. London: Sage.
A government agency or other technical document	Royal Society for Public Health (2018) *Health on the High Street: Running on Empty*. London: RSPH.
A journal article	Myatt, L.B, Scrimnshaw, M.D. and Lester, J.N. (2003). 'Public perceptions and attitudes towards a forthcoming managed realignment scheme: Freiston Shore, Lincolnshire, UK'. *Ocean and Coastal Management 46*. 565–582.
A magazine or newspaper article	Holmes, D. (2018) 'Geographical Skills. How to use qualitative data: researching place with interviews and social media'. *Geography Review 32* (2), 38–41.
An internet source	Simon Read and Tom Espiner (2018). 'VW emissions targets are too severe'. *BBC Business News*, available at: www.bbc.co.uk/news/live/business-47524588 *With websites it is crucial that you keep a note of the date accessed as the site may change over time.*

You might also want to reference within the text of an NEA. That might look something like this (note how the authors initials are not usually used in this convention):

> Research on coastal process in the 2000s included work in the UK by Massselink (2003) and in Australia by Hughes (2003) …

You should also provide a list of references at the end of the NEA, arranged in alphabetical order:

> Aagaard, T. and Masselink, G., 1999. Chapter 4: The Surf Zone. In A.D. Short (editor), *Beach Morphodynamics*, Wiley and Sons, London 72–118.
>
> Boggs, S., 1995. *Principles of Sedimentology and Stratigraphy (2nd Edition)*. Prentice Hall, New Jersey.
>
> Committee on Climate Change (2018) *Managing the coast in a changing climate*. London: Committee on Climate Change.

2 Footnote or endnote system

The alternative footnote style involves the use of numbered references in the text and a list of corresponding numbered references, usually at the bottom of the page. With word processing software it is very easy to insert the number in the text and the reference at the bottom of the page (or at the end of the work if you are using endnotes). The software will also ensure that the footnotes are in numerical order. The accrual numbers for the references are given either in brackets or as superscript, for example[1].

> In an early report, Dillion *et al.* (2006)[5] suggest that students remember fieldwork and outdoor visits for many years. Research from 1997 found that 96% of a group (128 children and adults) could recall field trips taken during their early years at school. However, simply recalling a visit does not mean that it was an effective learning experience or that the time could not be more usefully spent in the classroom. Several studies highlight the importance of carefully designed learning activities. Research by Ballantyne and Packer[6] in 2002 found that the use of worksheets, notetaking and reports were all unpopular with students and did not appear to contribute greatly to environmental learning. They suggest that touching and interacting with wildlife is a more effective strategy.
>
> ---
> [5]Dillion *et al.* (2006) 'The Value of outdoor learning: evidence from research in the UK and elsewhere.' *The School Science Review*
>
> [6]Ballantyne, R. and Packer, J. (2002) 'Nature-based excursions: school students' perceptions of learning in natural environment.' *International Research in Geographical and Environmental Education*

It may be a good idea to use a system such as *Onenote* or *Evernote* as a way of archiving the original copies of articles and internet pages that might be used either as part of a bibliography or reference list.

Preparing a bibliography

A bibliography provides a record of all the literature you have referred to when doing your research, including primary sources, secondary sources, articles and any visual or audio materials.

> ### Primary sources
> Primary sources are direct evidence. Examples of primary sources include statistical data obtained from experiments, audio and video recordings of interviews or people giving accounts of events, blogs and forums.
>
> ### Secondary sources
> Secondary sources are those materials which analyse and evaluate someone else's original research (primary sources). Examples of secondary sources include newspaper, magazine or journal articles and books.

These works are not simply lumped together in a list, but are normally organised into sections with headings:
- Primary sources
- Secondary sources
- Articles
- Other materials

Your school, college or teacher may have a policy on the referencing system they want you to use and it would be worth checking to see before following the guidance given here. AQA, at the time of writing, has no set requirement of the system that is used, but consistency is the key.

Excessive referencing

A certain amount of judgement must be used when a large number of references are made to the same, single source. For the reader, it becomes tedious when the same repeated reference appears too much. One way to avoid this is to state that a large selection of your literature review comes from a single author, or a single source document. For example, if you are summarising MacCannell's work on tourism, rather than have a large number of individual references, you may want to create a separate section in the NEA, for example:

> **The work of Dean MacCannell**
>
> This section of the literature review summarises MacCannell's (1976) influential work, *The Tourist; A New Theory of the Leisure Class* in which the author provides a classic analysis of travel and sightseeing …

Research skills and plagiarism

Plagiarism is pretending that someone else's work is your own. It is very rare for students to download a ready-made essay from a website and say they wrote it, but sites which offer projects for a fee should be avoided. It is also very rare that students will get someone else to write their coursework, as this is dishonest and also teachers would see that the work was clearly not that of someone they had taught and whose work they knew. However, a great deal of material is available on websites and can be easily copied and pasted. Textbooks, articles and specialist studies can be photocopied or photographed. Many students like to build up a body of material like this, but it is very important when researching that there is a distinction between work read, consulted, used and made part of your own work, as opposed to material that is just inserted without any acknowledgement. That sort of plagiarism is not so obvious as the use of someone else's whole work but it is just as important to avoid it.

The use of books and articles

When research is undertaken it is vital for you to make notes which get to the heart of ideas, arguments and supporting factual information. If a particular concept or assumption is used then it is often helpful to quote briefly, but long extracts should be avoided and the write-up of the NEA should not consist of a series of quotations. When a direct quote is made then this should be referenced or footnoted (see page 67–68). If an idea is taken from a geography book then it should also be acknowledged.

However, it would obviously not be appropriate to acknowledge where you obtained every fact. If you are starting without previous knowledge, let's say knowledge of coastal erosion processes, and you read a book or article which tells you about attrition and abrasion, then you would not need to acknowledge where you got this information as it is common knowledge. However, if you decided to take more precise knowledge about a coastal process (for example, linked to measured rates of cliff recession in an area) from a book or article then this should be acknowledged. You are using someone's research and ideas and they will be happy that their work is being read and used – but it is common academic courtesy to acknowledge that it is their idea that you are using and not your own. The marks at the higher levels come from expressing a clear and consistent view and argument – it is very difficult to do this if you are simply attempting to put someone else's views in your own words.

The use of primary evidence

To introduce an idea, opinion or explanation from a primary source is not quite the same as using one from a secondary source, but it is necessary to reference the evidence. For example, by using a footnote you are showing where the argument came from and making clear that the interpretation is not yours but is taken from a source.

The use of internet material

This material can usually be copied very easily. It is extremely important that sections of work are not lifted unacknowledged from online encyclopedias, information pages or articles. As with books, it is strongly recommended that notes are taken from the material and that arguments and ideas are acknowledged. If there is a quotation, then the web address and the date of accessing the site should be included.

The use of secondary evidence

Many books outline the geography of the topic they are dealing with. This is often very helpful in coursework, but students should not give the impression that they have read and used all the sources mentioned and should acknowledge that they are using a digest of different views by an author or authors, not that they have consulted the authors or geographers mentioned when they have not.

My progress

Use the following checklist to ensure that you have taken all relevant information from each source you consider.

Title, author, date of publication and publisher	
Summary of key ideas or concepts	
References to other works that can be followed up	
Other research that supports it?	
Other research that contradicts it?	
How far does it offer any empirical evidence or data that can be tested against my own fieldwork and research?	

Self-assessment

As this coursework is an independent piece of work, the amount of help your teacher can give you is limited by the rules set down by the body that regulates all A-level examinations in Geography. This means that your teacher cannot offer direct feedback and can only give general guidance about titles rather than make specific suggestions. Your work is going to be marked against set criteria (see pages 8–11), so it is important that you try to apply the mark scheme yourself before you hand in the work in its final form for marking.

If you get used to making use of the mark scheme as you are producing the coursework, then this should help you to be sure that you are on the right lines.

Looking at your investigation, you have to ask yourself whether it meets the requirements set down by AQA. You can do this by filling out Table 25 and referring to the checklist on these pages. Don't wait to the end to fill this in. Once you start writing, refer to it as much as possible. Use this against some of the checklists on pages 8–11.

Table 25

Key questions to ask	Yes/mostly/partly/no	What do I need to do to meet it?
Have I got the necessary background knowledge about the topic to undertake an investigation at this level?		
Have I completed an appropriate literature review for the topic?		
Have I used an appropriate idea, model or concept?		
Have I designed a methodology which is manageable and appropriate?		
Have I used detailed knowledge?		
Have I recognised the ethical dimensions of the study?		
Have I explained, assessed and made links, rather than included a lot of description?		
Have I related everything in my coursework to the original focus?		
Have I explained what impacts reliability and accuracy might have on outcomes?		
Have I reached a reasoned conclusion on which of the chosen works presents the most persuasive argument?		
Have I made an overall judgement at the end, and made links to wider geographical understanding?		
Have I correctly referenced and paginated the work and made a check on all spelling and grammar?		

Reflecting on a 'quality product'

The lessons learnt from marking past geography projects reveal a number of important considerations. Titles and introductions are an area where things can go wrong. So, check that:

- Your title is focused on small areas or a place which is safe and manageable.
- You can see a clear purpose in what you are doing (and you understand the concepts and processes linked to this).
- You have carefully considered either two or three key questions or sub-hypotheses to provide a framework for your investigation.

There are several other ideas that also need consideration in relation to achieving a quality end product.

1 Have you used 'pilots' or trials of different-sized areas or techniques?

2 Have you done the research and write-up about the study area, its context, processes and so on before carrying out any fieldwork?

3 Have you used authoritative secondary sources in an integrated way (often at the start of the investigation)?

4 Have you ensured that your data were achievable, substantial, and reliably sampled?

Read the statements in Table 26. Use a circle for each line to show where you think your work is currently positioned. You do need to be brutally honest here!

Table 26

The best work ...	through to	... the less good work
Genuinely individual work		More or less what I have done before (feels 'safe')
A geographical focus based on wide research and understanding		Focus based around what is available in the textbook
Focus is linked to specification		Focus barely linked to specification (and dangerously non-geographic)
Achievable title/questions		Title too big and not really answerable
Focus strongly on place context and relevant processes		Too much on place (historical) and processes not understood
High quality research in terms of primary data		Weak quality research in collecting primary data
Sampling understood together with sample size		Small sample sizes, not based on valid reasoning
Evidenced conclusions		Generalised conclusions
Genuinely reflective and accepting of shortcomings in terms of impacts on conclusions		Unquestioning evaluation ('I have certainly proved that')

Research logs

It is good practice to keep a resource record which will act as a log of the literature research you have carried out (we have already looked at this in Section 2). It is also useful to keep a log of your initial research and ideas. The log below has been filled in to give you an idea of what to do. As you work on your coursework you can fill out the blank versions on the subsequent pages in this workbook.

Getting started – deciding on a broad topic and thinking about the geographical concept

Initial ideas for NEA	Why I chose it	Initial search for resources
June Problems on the high street? The 'Amazon effect' Change in the loss of identity on the high street? Changes in people's shopping habits and attitudes? A sustainable future for the high street? Crime in the high street?	I found this interesting in my A-level Paper 2 course but want to know about this in more depth. We did look at some theories but only for a few lessons. There is a lot of material available and different views to look at and it is very topical. Decided against the coast since it seemed like lots of people were doing that and it might be difficult for me to show individuality.	Did an internet search on the problems on UK high streets and changes and started to look at A-level texts. Found a university book with a good section on causes. Found websites on the debate, e.g. Mary Portas' 'Save the High Street'.

Refining the title and working-up geographical ideas

Working title	Explanation and notes	Supplementary reading	Reflection
September Does the changing nature of the high street change its identity?	Shops and buildings are part of the place identity. Loss of shops will mean a more residential function, changing the high street.	I found sources online, for example 'Town centres could become ghost towns' from the BBC, and then found a linked article, 'The Grimsey Review 2'.	In a discussion, my tutor suggested that 'why' might lead to a list and that I should make sure that my title leads clearly to a discussion.
October Does the centre of Shrewsbury offer a healthy environment in terms of perception of crime?	Changed because I needed to be more specific in a location and make it more workable in relation to health (I had found lots of research on this). The aspect of crime seemed measurable.	I found some work on 'Healthy High Streets' and the sort of indicators that could be measured.	Title approved, but tutor agreed it could be a work in progress. My tutor was worried about how big my project could become.

Initial design and methodology research

Date	Resource(s)	Key ideas	Evaluation
October	'Healthy High Streets: good place making in an urban setting.' Public Health England and Institute of Health Equity (2018). 'Public perceptions of crime in England and Wales: year ending March 2016.' ONS	Crime is an aspect of health on the high street, along with many other variables (too many for me to measure). Can I map perception of crime and then bring into a broader discussion around health?	Need to find more information to work out sample size and design, e.g. using Shrewsbury latest census data so that sample can be stratified to reduce bias. Need to think about photography to record primary evidence as well.

Example literature record/review

Resources used	Page/web reference	Student comments	Student date(s) when accessed
Holmes, D. (2018) 'Geographical Skills. How to use qualitative data: researching place with interviews and social media.' *Geography Review* 32 (2).	Pages 38–41	Outlines fieldwork methodologies in terms of collecting primary data using archive materials. I decided to use parts of the methodology, and also explore the use of social media as a way of collecting additional evidence to support my focus.	October
Royal Society for Public Health (2018). 'Health on the High Street: Running on Empty.' London: RSPH	Pages 10–29	This study looks to show how changes on the high street are causing health problems, e.g. unhealthy shops and activities. It also has a fieldwork methodology. The background concepts in here are useful and will allow me to contextualise my study and show wider linkages.	October

Broad topic and thinking about the geographical concept

Initial ideas for NEA	Why I chose it	Initial search for resources

Refining the title and working-up geographical ideas

Working title	Explanation and notes	Supplementary reading	Reflection

Working title	Explanation and notes	Supplementary reading	Reflection

Initial design and methodology research

Date	Resource(s)	Key ideas	Evaluation

My literature record/review

Resources used	Page/web reference	Student comments	Student date(s) when accessed

Schedule planner

Final due date:			
Stage	Approximate number of hours required	Schedule start and finish date	Completed?
Introduction and preliminary research			
Methods of field investigation			
Methods of critical analysis			
Conclusions, evaluation and presentation			
Additional checks and proof reading			

Glossary

Accuracy The closeness of measurements of a quantity to that quantity's actual (true or real) value. In fieldwork this can be very difficult to determine so the word needs to be used with caution.

Analysis The stage in the report where you write about what you have found, provide explanations, make links between findings and so on. Evidence is always required.

Annotation The process of adding detailed notes, processes and explanations to photographs and images.

Anomaly A different result or deviant from the general pattern or trend. It may be genuine or it may have bene introduced as a result of measurement error.

Bias (sampling) An inclination or prejudice towards, or against, a specific finding or outcome. Bias will create distortion in a statistical result.

Cartographic The process of drawing maps which are often included as part of the data presentation or representation. Maps may also be overlaid with graphs and other diagrams.

Conclusion The summary of what you have found – the final finishing section. This must be based around evidence you have provided in the rest of the work.

Empirical data Evidence and information acquired by observation or experimentation. This can be used to make comparisons.

Evaluation A reflective process, saying what was good or bad, commenting on the reliability of results. In the NEA, evaluation should also consider reliability and accuracy, in other words how much you trust your results, and whether you have actually answered the question you set out to answer.

Fieldwork Usually associated with going outside and collecting data about people, places and environments. Some fieldwork may be more research based or take place inside buildings.

GIS Geographical Information System – a geo-spatial tool to represent points, places and areas on a digital map.

Graphical (representation) Using different graphs to show information, often data collected from fieldwork.

Harvard A commonly used referencing system to incorporate other people's quotes, findings and ideas into other work, in order to support and validate their conclusions. It is this system that is often used when writing for an academic purpose.

Hypothesis A testable idea in the form of a statement (not a question).

Interview Longer, open-ended style of questioning someone; really like a conversation.

Introduction The start of the report, setting up what you intend to do and giving background information.

Median Divides the data into two halves; the median is the middle value (which may be different to the mean).

Mode The most frequently occurring number in a series of numbers.

Opportunistic (sampling) This sampling frame is one of convenience and is associated with questionnaires. Typically it uses people from a target population available at the time and willing to take part as respondents.

Pilot survey A practice or trial survey to ensure that data collected will be appropriate to the question, for example testing locations and methods.

Plagiarism Passing off the work of others as your own. Referencing other works is always required.

Population (sample) The broad group of people, feature, items, etc. from which you take a sample and then generalise the results.

Primary data Data that you have collected yourself, first hand – it may come from other sources if it remains 'raw' and unprocessed.

Qualitative data Information which is subjective or does not have any number(s) such as a photograph, sketch map or an interview transcript.

Quality of life A broad idea of how pleasant or agreeable an area might be in terms of housing, schools, environment, and so on. It generally relates to people – they have a high or low quality of life.

Quantitative data Data which contains numbers and figures such as the number of pedestrians, size of sediments, closed questionnaire response data, and so on.

Question A geographical question that might be asked at the beginning of a fieldwork investigation.

Questionnaire Usually a short interview with a single respondent where there are lots of questions and factual, numbered responses which produces quantitative data.

Random error These cause results to be spread about the true value. The true value is the correct or right answer to the thing you are trying to measure. The effect of random errors can be reduced by taking more measurements.

Range The difference between the highest and lowest values in a set of data.

References Details of any published work or research you have used as part of your work. This forms part of the research literature.

Reliability The consistency or reproducibility of a measurement. A measurement is said to have a high reliability if it produces consistent results under consistent conditions.

Report The work that you will hand in for the fieldwork investigation, i.e. the write-up. It will be marked by your teacher and may be selected for external moderation by the awarding body. This is to double check and confirm the original marking.

Risk assessment The procedure of identifying and subsequently managing potential risks.

Sampling An accepted shortcut, in other words, a way of getting data for your study without collecting everything about a population. There are several different types of sampling strategy, procedure or sample frame, which are all aspects of sample design.

Secondary data Data that you got from someone else or another organisation that is often in a written-up form.

Stratified (sampling) A sampling procedure when you take into account known information about the characteristics of a sub-population and adjust your sampling to reduce bias.

Systematic (sampling) A sampling procedure that uses a regular sampling frame; for example, collecting data every ten minutes, or every seventh person.

Systematic error These cause results to differ from the true value by a consistent amount each time the measurement is made. For example, a student uses weighing scales which have not been zeroed, so all the results are 10 g too high.

Theory A geographical idea or concept that may underpin the reason for your aims or questions.

Transect A line along which you carry out sampling, such as a road or river.

Variability A lack of consistency or fixed pattern; liable to vary or change.

The Publishers would like to thank the following for permission to reproduce copyright material.

Photo credits

p.31, **p.34**, **p.35**, **p.45** and **p.54** © David Holmes

Text permissions

p.24 Table 10: The Local Data Company (LDC); **p.24 Figure 8**: Frederick O'Brien, The Hipster Index: Brighton Pips Portland to Global Top Spot, Move Hub, 19 Apr 2018; **p.26 Figure 9** and **p.29 Figure 12**: L.B. Myatt, M.D. Scrimshaw, J.N. Lester, Public perceptions and attitudes towards a forthcoming managed realignment scheme: Freiston Shore, Lincolnshire, UK, 2003 Elsevier Science Ltd. All rights reserved. doi:10.1016/S0964-5691(03)00035-8; **p.26 Table 1**: Reprinted with the permission from Dr. Jennifer Ferreira; **p.29 Figure 11**: Larkin Gowen, Tourism and Leisure Business Survey 2018 © 2019 Business Tourism & Leisure Survey.

Acknowledgements

Every effort has been made to trace all copyright holders, but if any have been inadvertently overlooked, the Publishers will be pleased to make the necessary arrangements at the first opportunity.

Although every effort has been made to ensure that website addresses are correct at time of going to press, Hodder Education cannot be held responsible for the content of any website mentioned in this book. It is sometimes possible to find a relocated web page by typing in the address of the home page for a website in the URL window of your browser.

Hachette UK's policy is to use papers that are natural, renewable and recyclable products and made from wood grown in well-managed forests and other controlled sources. The logging and manufacturing processes are expected to conform to the environmental regulations of the country of origin.

Orders: please contact Hachette UK Distribution, Hely Hutchinson Centre, Milton Road, Didcot, Oxfordshire, OX11 7HH. Telephone: +44 (0)1235 827827. Email education@hachette.co.uk Lines are open from 9 a.m. to 5 p.m., Monday to Friday. You can also order through our website: www.hoddereducation.co.uk

ISBN: 978 1 5104 6877 1

© David Holmes 2019

First published in 2019 by
Hodder Education,
An Hachette UK Company
Carmelite House
50 Victoria Embankment
London EC4Y 0DZ

www.hoddereducation.co.uk

Impression number 10 9 8 7 6 5 4 3 2

Year   2023 2022 2021

Cover photo © dikobrazik - stock.adobe.com

Illustrations by Aptara Inc.

Typeset in Bliss light 11/13 pts by Aptara Inc.

Printed in UK.

A catalogue record for this title is available from the British Library.

HODDER EDUCATION

t: 01235 827827
e: education@hachette.co.uk
w: hoddereducation.co.uk

ISBN 978-1-5104-6877-1

9 781510 468771

MIX
Paper from responsible sources
FSC™ C104740